How to Start a

VoIP Business

A Six-Stage Guide to
Becoming a VoIP Service Provider

Vilius Stanislovaitis

How to Start a VoIP Business:
A Six-Stage Guide to Becoming a VoIP Service Provider
Vilius Stanislovaitis

Copyeditor: Val Dumond and James Millington
Cover design: Ryan Lause
Diagrams design: Michael Cissell
Book Interior: István Szabó, Ifj.

If you would like to do any of the above, please seek permission first by contacting me at
http://www.runvoipbusiness.com

Published in Lithuania by Vilius Stanislovaitis
ISBN 978-609-408-830-8
Version 1.0

Table of Contents

The Story
Behind This Book

The story of this book began in 2009 at Kolmisoft, a company that specializes in providing billing and routing platforms to manage Voice over IP (VoIP) businesses. Some of the company's customers were existing VoIP service providers. Others were entrepreneurs who had launched their businesses based on VoIP technology for the first time.

Some of the startups wanted to launch a calling card business by offering cheap international calls to the ethnic community within their market. However, starting a business from scratch was frustrating for the entrepreneurs. They needed to work through a number of specific tasks, such as reviewing the telecommunications regulations in their market, performing a client and competitor analysis, establishing a company, finding reliable suppliers and partners, organizing sales and marketing, choosing the software and hardware solutions, and many more.

Going through all those procedures for the first time without any guidance made startups feel stressed rather than excited about the new business opportunity. Some of them gave up in the middle of the process because of limited time and energy. One of the main reasons for the early failures was that they couldn't find a trusted resource that would supply a step-by-step guide to starting and running a prepaid calling card service.

After a detailed market observation and analysis, the

white paper *How to Start Calling Card Business* was released as a solution. The white paper enabled entrepreneurs to break a complex, overwhelming project into smaller manageable tasks, making it easy to start with the first one and to move on from there.

The demand for such a guide encouraged the development of the primary idea of that white paper into this informative yet simple book with practical examples, in order to explain the launching of different types of VoIP business.

Who Should
Read This Book

The book is written in very simple, non-technical language, making it easy for most readers to follow. There are four main groups of people whom the book serves. You'll most likely recognize yourself in at least one of them.

Startups and Entrepreneurs
These are people new to the telecom industry who are planning to launch a VoIP business in the near future. One of the key challenges for entrepreneurs is to create a realistic business plan and follow it. However, that is not so easy when you are starting a VoIP business from scratch. Putting different pieces together is a frustrating and time-consuming task. A more simple and convenient way is to develop a plan according to this guide, which is based on hundreds of real business cases. The book will be your "virtual mentor" and will lead you through the main stages of starting your own VoIP business step-by-step. You'll be able to learn from the success stories and mistakes of those people who have already gone through this process before you. Having a complete guide saves time and will keep you focused and productive.

Business Owners
There is no doubt that business owners and co-founders of VoIP companies will pick up this book excitedly. Not be-

cause they need a guide to start something new, but because this book is a story about them. All business owners were startups at some point and have gone through similar stages to those described here. This book reflects those ups and downs. Company founders familiar with this project were eager for this book, in order to give it as a gift to employees and clients, helping them to see the whole picture behind a VoIP business.

Technical People

The majority of these people are current or former employees of telecommunications companies with experience at the technical and operational level. One of the key problems for technical people is that they sometimes pay too much attention to the technology instead of focusing on the revenue-generating activities. In some cases, technicians spent a few years building a perfect VoIP platform that was ideally integrated with their website; and a majority of the processes where fully automated. The only issue was that they completely forgot whom they built it for; there were not enough customer attraction activities. As a result, most businesses such as these failed.

This book helps you to understand both the technical and commercial aspects of a VoIP business. You will learn what to focus on, which tasks to delegate or automate, and what to cross off your list.

Curious People

If you did not recognize yourself in any of the above profiles, but you are a curious person who'd like to learn more about VoIP businesses, you'll enjoy this book. The fact is that a lot of people are familiar with the VoIP concept, but only a few understand what is happening inside the VoIP provider's environment.

This book offers solutions to something that is a mystery for many people—how to organize processes to generate revenue from VoIP services.

How to
Read This Book

A VoIP business is like a puzzle which consists of many small pieces that should be connected in the right way to get good results. Everyone can complete a puzzle; the only difference is: how fast? To make the complex simple, all of those pieces have been placed into six informative, easily-read chapters. The section titles are defined clearly so that they can be skipped by those who are already familiar with the material.

Chapter 1. Understanding Telephony
This is an introduction for those who are new to telephony. This part explains the evolution of the telecommunications industry and how it has changed over the decades. You'll also learn the benefits and value of VoIP and how it differs from traditional telephony.

Chapter 2. Choosing a VoIP Service
There are many services that can be provided by using VoIP technology. This part reviews the special aspects of various business models and services.

Chapter 3. Deciding on a Business Model
There are two VoIP business levels that you can choose from: reseller and service provider. This part will help you decide on the right business model according to your strengths. Besides this, you will learn how to define your target audience.

Chapter 4. Softswitch Selection

Softswitch is the key platform that connects calls between clients and providers. This part explains how to build a solid infrastructure and what components are needed to run different types of VoIP services. The content is not too technical; all terminologies are explained, so it is easily understandable.

Chapter 5. Finding Partners and Suppliers

To run a VoIP business, you'll need to select multiple partners who will provide different resources for you. This chapter explains what type of partners and suppliers you need in a VoIP business.

Chapter 6. Launching Your Service

You're ready to go. This chapter offers ways to organize sales and marketing, define your ideal client, and begin a business with minimal resources. You will also learn the most common operations handled by VoIP providers daily.

Glossary

At the end of this book, you will find a glossary, which explains the most common terms, used in a VoIP business.

NOTE FROM THE AUTHOR: Before you read further, a note regarding expectations. This book is a guide to starting a VoIP business, but it has nothing to do with self-motivation or topics like "how to become a millionaire." Most probably you will not become a millionaire, but you'll enrich your knowledge and understanding of a VoIP business.

So relax, grab a cup of coffee, and let's get this journey started!

Chapter 1.
Understanding Telephony

Telephony is a domain of telecommunications that refers to transmissions of voice, video and data using electronic media (wire, radio, or fiber optics). Telephony has brought great benefits to our lives and now serves an increasing role in the business environment.

Voice communications have become an essential part of company practices, improving the efficiency of internal and external communication and helping to maintain and develop better relations with clients and partners. Moreover, telephony as a business service saves time and money for both residential and business users while still helping to earn money for telecom service providers.

This chapter leads you through the evolution of the telecom industry, compares the different ways used to transmit voices, and focuses on the value of thriving VoIP technology.

Evolution of Telephony

People have always been looking for the most efficient way to communicate remotely. The history of remote communication began with smoke signals. Later remote communication expanded to include mail, pigeon post, and maritime flags.

But probably the greatest innovation was achieved when communication could become electrical. The story of telephony began with the invention of the electrical telegraph and, later, the telephone. It was the radical invention of those times, and it changed forever the way people communicate with each other.

At first, telephone communication was point-to-point; the voice was transmitted over analog signals. To establish a call between two people, telephones had to be interconnected by wires. But instead of linking all telephones to each other, a more efficient solution was employed. Phones were connected to a central telephone network mechanism called a switchboard.

The switchboard was a telecommunications system used to establish telephone calls between subscribers, or between other telephone exchanges. They were manually controlled by switchboard operators who were responsible for connecting calls between two parties. Mechanical switchboards marked the beginning of the public switched telephone network (PSTN) in the United States in 1878. The PSTN was an aggregate of the telephone exchanges, phones, and interconnecting facilities that were used to provide phone services to the public.

Even though switchboards enabled more efficient communication, telephone services had one big problem. It was possible to send only one signal over a wire. The price of such connections was very high as each call had to be established manually; a pair of copper wires could not be used for more than one conversation at a time. The investment to set up a telephone network was therefore huge, and only large corporations could afford it.

But things changed a lot at the end of the 19th century when dial telephones and automatic telephone exchanges were invented. Manual switchboards were gradually replaced, and communication became faster and more affordable. As a result, wired networks expanded as the demand for telephone services rose.

By the middle of the 20th century, microwave communications appeared, allowing signals to be transmitted through the air in areas that could not be covered by wires. In addition to this, Bell Laboratories invented the transistor, which then lead to the development of a computer-based electronic switch (the current version of it is called a "softswitch," short for "software switch"). (See the "Softswitch Selection" chapter for more information on softswitches.)

These upgrades established the start of modern electronics. Development of new technologies gradually transformed analog signals to digital (a process sometimes called "digitization"). Digital telephony turned signaling and voice into a stream of bytes, making it possible to send multiple signals over the same circuit. This increased the call capacity per wire, reduced the cost of the network, and improved voice quality.

Digital technology and the advent of the personal computer in the 1980s opened new opportunities in telephony, making it possible to use a computer to initiate phone calls and manage the calling process. Computer networks became similar to telecommunications networks, though the network performance was not initially sufficient to carry a good quality of voice.

In the mid-1990s, computer network performance improved to a point where it became possible to send a stream of media information across a network connection. A media stream was chopped up into segments, which were wrapped in an addressing envelope. Due to their origin, such computer-based connections were referred to as "packet-switched."

The biggest challenge was to send a flood of these packets between two endpoints while ensuring two conditions. First, that packets should arrive in the same order in which they were sent and in an appropriate timeframe. Second, that none of the packets could be lost. Once those conditions could be reliably reached, VoIP became a well-known alternative to standard telephony.

Traditional Networks

From the global perspective, the telecommunications world consists of many semi-autonomous networks which are operated by different types of providers. This section reviews the most popular telephone networks and the operators that control those networks.

PSTN

The PSTN, which appeared at the end of the 19th century, was based on circuit-switching technology that used a dedicated channel (circuit) for the duration of the voice transmission. Historically it was composed of a hierarchical structure; each phone subscriber had a direct connection to the PSTN through the local exchange that served end-users in a small area. The telecom operator which provided the local telephone service was called an "incumbent local exchange carrier" (ILEC or LEC). Bell Atlantic was one of the first LECs in the United States to provide all types of calls: local, national, and international.

Analog and Digital Networks

The PSTN originally carried only analog signals, but today it is almost entirely digital and includes mobile phones in addition to fixed-line phones. Only the oldest parts of the PSTN still use analog technology for the so-called "last mile" connections. Such connections are referred to as the "plain old telephone service" (POTS). It is the basic analog voice service, traditionally transmitted over copper wires to individual homes and other end-users.

The availability of digital connections has increased through services like cable, DSL (digital subscriber line), and ISDN (integrated services digital network). ISDN allows multiple voice and data signals to be transmitted over a single telephone line, and it comes in two flavors: BRI (basic rate interface) and PRI (primary rate interface).

BRI is referred to as "2B+D" and consists of two B channels (which carry data, voice, and other services) and one D channel (which carries control and signaling information). BRI is mostly used by home users and small enterprises.

PRI is referred to as "23B+D" (or "30B+D") and consists of 23 B channels and one D channel using a T1 line (in the United States, Canada, and Japan) or 30 B channels and 1 D channel using an E1 line (in Europe and other countries). E1 and T1 lines are widely used by Internet service providers (ISPs) for connection to the Internet, and by Internet telephony service providers (ITSPs) for interconnection with operators.

Mobile Networks

A mobile network (also called "cellular") is a wireless network distributed over areas, called "cells." For some people, the term "cellular" sounds quite old, since lately everyone is used to hearing the terms LTE (long term evolution), 3G (third generation), 4G (fourth generation), and 5G (fifth generation). The basic difference between them is the speed of the network, which increases with each generation. Even though the deployment of 4G networks has not yet been completed, development of the fifth generation has been started, intending to be commercialized by 2020.

Mobile networks are deployed over most inhabited land, and the network infrastructure is growing at a tremendous rate. There is no doubt that the world has gone mobile; more than 85% of the population has access to a mobile phone. Even though mobile networks were created to transmit voice services, the current usage of mobiles has changed significantly. Mobile users listen to music, watch videos, read books, play games, communicate in social networks, and surf the Internet on their mobile phones regularly. The need for mobile data is increasing, so the industry is continually working on speeding up the development of data networks.

Voice over IP

VoIP is probably the best-known service that uses packet switching to transmit calls. It is not a brand new technology, but many factors in the past have slowed the anticipated growth of IP telephony. The first VoIP call attempt in the 1990s was a huge step forward in telephony, but was still terrible in quality. The reason was not the technology itself, but the Internet bandwidth which was not sufficient for carrying voice packets.

However, things have changed a lot since then. Now broadband capacity has increased a few times and allows more data packets to travel through the Internet channel.

Moreover, transmission is also done much faster. A well-planned VoIP network can offer call quality and reliability equal to mobile or landline calls.

Protocols and Codecs

Different from circuit-switched technology, packet-switched networks do not require a dedicated path through the network. The caller's voice is digitally encoded into fragmented data packets that are transmitted over the Internet connection. At the destination of the call, packets are reassembled in their original order, and the data is decoded to voice.

Encoding and decoding are done by codecs that convert an analog voice signal to the digital version and vice versa. There are many different codecs that are used in VoIP: G.729, G.729A, G.729B, G.729AB, G.723.1, G.711, and GSM. Codecs play a big role in voice quality (to be reviewed in more detail later in this section).

The protocol is a set of rules used by the end points (e.g. two phones) when they communicate to each other to establish a connection. There are a variety of proprietary protocols (SKINNY by Cisco, IAX2 by Digium) and open protocols (H.323, MGCP) used in VoIP, but the most popular are Session Initiation Protocol (SIP) and Real-Time Protocol (RTP). SIP is used as a signaling protocol to set up a call; it works in two phases. First, it establishes the call by allowing two phones to "talk" to each other. Then it starts the control of media transfer between two phones. The actual data transmission uses RTP.

VoIP Quality

VoIP quality is one of the most common subjects of discussion; many people doubt whether VoIP can have the same quality level as traditional telephony. There is no single answer to this question, but there are few factors that influence voice quality.

Jitter

A timing variation in packet arrival, which can occur because of timing drift, route changes, or network congestion. Jitter below 25 milliseconds (ms) is acceptable, but if it is between 25-100 ms, a jitter buffer is recommended. This is a shared data location that collects and temporarily stores arriving packets before sending them to the voice processor in evenly-spaced intervals. This process minimizes the delay variations and makes the connection clear with very little sound distortion. Jitter buffers can be dynamic (software-based, configured by the network administrator) or static (hardware-based, configured by the manufacturer).

Latency (Delay)

The time it takes a conversation to travel from the speaker to the listener. Latency below 150 ms is acceptable, but if it is higher than that, the call recipient hears an echo. Delay can happen if, during a call, another user on the same network downloads a large file. Without packet prioritization, call quality could be degraded. The solution for avoiding latency is to give voice traffic the highest priority. It ensures sufficient bandwidth to carry voice packets and prevents delay-related issues.

Packet Loss

May occur when connection speeds are compromised by temporary Internet provider problems (network congestion or heavy bandwidth usages such as online gaming or the downloading of large files). A mere 10% of packet loss can cause a serious impact on voice quality. The solution is the same as in latency—prioritize voice traffic.

Echo

The experience when you hear your own voice coming back through the telephone, sometimes with a significant delay. Perhaps you've experienced it at least once – varying from being unpleasant to being downright annoying. In most cases, conversations with high echo amplitude are quickly abandoned by participants. Similarly to jitter, an echo interval below 25 ms goes unnoticed, but if it exceeds this limit, it will become audible to the end-user. Many factors influence the delay and echo: audio codecs, transmission lines, and network devices, but the main source of this quality problem in a VoIP provider's environment is usually the interconnection point between the VoIP device and the PSTN. As a solution, it is recommended to use devices that support echo cancellation functionality.

Devices

Refers to phones, routers, network adapters and telephone systems. It is recommended that you do your homework (read specific forums, reviews, ask for recommendations, discuss with suppliers and users) before investing in VoIP equipment. Many new companies make their purchase decisions based on price, if product characteristics appear to be the same. In reality, characteristics do not determine every-thing, as many manufacturers copy the technology of well-known brands and try to sell their copy at a lower rate. In the majority of cases, there is a reason for their lower price. It usually represents equipment of lower quality (maybe not in the product itself, but in technical support, the replacement policy, or the guarantee). Moreover, it is important to take into consideration the interoperability between each hard-ware component and the general network topology. The hardware you choose may be the best in the world, but you will still have problems if it is not compatible with other ele-ments within your network.

Codecs

The most commonly used are the G.711 and G.729 codecs which differ from each other in their sound quality and bandwidth consumption. (The G.711 codec requires more bandwidth, so its voice quality is better than G.729.) It is also important to note that the majority of free applications and cheap devices are not using the licensed version of the G.729 codec, which can affect customer perception in the narrow bandwidth networks. If you plan to provide VoIP services in less developed countries where the Internet connection is poor and expensive, it is recommended that you use the bandwidth optimizer and codec which consume less band-width. On the other hand, if you serve clients who have broadband access, it is worth mentioning the G.722 codec, which provides improved speech quality due to a wider au-dio bandwidth, although it is not as widespread because of high bandwidth requirements.

Bandwidth

This was the core reason for bad voice quality when VoIP first appeared. Nowadays, the broadband network allows the allocation of sufficient connectivity resources for real-time voice traffic, so bandwidth rarely becomes a problem. It's an issue only in rural areas and less developed countries where connectivity is delivered by Wi-Fi, WiMAX, satellites or balloons. In this case, to ensure better voice quality and reduce the expense of Internet connectivity (which is relatively high in some countries), it is recommended you use VoIP bandwidth optimization solutions (also called "bandwidth optimizers"). They reduce bandwidth consumption by up to 50-80% without compromising on call quality. Enhanced bandwidth use leads to improved quality performance and higher call capacity through the same channel.

VoIP Benefits

These are the main benefits of VoIP:

- **Free**. Calls inside the network (also called "on-net") are free. It is especially useful if you have remote offices in different cities or countries.
- **Cost savings**. First, VoIP channels are provided at a lower rate compared to traditional telephone lines. Second, there are significant cost savings on international long-distance calls.
- **Accessibility**. You can access your account and make VoIP calls from a web browser, IP phone, smartphone application, or computer program.
- **Portability**. You can easily pick up your phone and

move to another geographical location. VoIP works with the same account details as long as there is an Internet connection in your new location.

- **Value-added service.** All the functions of old telephone systems are available in VoIP, and many new features are introduced continuously. Usually, they come with no additional cost.
- **Up-to-date.** Traditional telephony providers are becoming outdated, meaning they will either lose their customer base due to non-competitive services or will streamline their operations and migrate to VoIP.

The above advantages are not relevant to everyone, so it is important to understand the value drivers for end-users and corporate clients. If asked to choose the customer's primary value-driver of VoIP in the end-user market, the answer would undoubtedly be cost savings (especially on long-distance calls). Of more importance to the corporate segment is to increase the productivity of individual employees as well as overall business efficiency by using advanced communication systems.

Looking at it from the entrepreneur's point of view, running a VoIP business is attractive because of a few factors: simplicity, low investment, scalable income, and the possibility to manage the operation from any part of the world.

Telecom Regulation

The telecom market was deregulated in 2000. Since then, new players (often named as competitive local exchange carriers (CLECs)) have been allowed to compete with the former monopoly operators (ILECs). The pace of deregulation was different in each region; some countries did not even consider this process, as revenue from the telecommunications sector was quite significant.

The process of deregulation has resulted in the need to create an institution to control all processes related to the telecom market. Currently, each country has a telecommunications regulatory body that governs the market of fixed and wireless telecoms, Internet transmission, broadcasting services, and VoIP.

Depending on the region, there are a few approaches to VoIP. It can be illegal, legal but unregulated, have light regulation (usually the transitional stage before making a decision to regulate) or regulated.

VoIP is illegal in those countries that still have a telecom monopoly. Such markets have restricted competition, and the dominant telecom operator is owned by the government or an organization associated with the government.

Monopoly-oriented governments usually strictly regulate the telecommunications market or even block VoIP to keep control of it and protect revenue. On the other hand, in a majority of developed countries, VoIP is legal. Such regions have strong competition and a mature market where the main discussion issues are related to emergency service support, user privacy policy, and number portability.

VoIP, as a technology in developed countries, can be treated differently. It can be regulated in the same way as traditional telephony (PSTN), considered as a computer-based information service, or as a unique service that has its particular peculiarities.

Geographical Regulation

Regulatory laws are being changed constantly by governments. To get up-to-date information on regulations in your own country, it is recommended that you contact your local authorities. This subsection reviews the approach to VoIP regulations in different regions.

An interesting correlation that reflects the regulation level in specific countries is the difference between Skype rates and interconnection rates applied by the local telecom operator. (An interconnection rate is the rate at which a provider agrees to terminate calls inside its network.) You can check the results in the table below:

Country	Skype rate	Interconnection rate	Percent difference
UK, USA, Canada	2.1 cents/min	2.1 cents/min	0%
Bolivia	11.4 cents/min	31.5 cents/min	276%
Nicaragua	21.2 cents/min	$1.97/min	928%
Cuba	71 cents/min	$19.74/min	2780%

A low percentage represents liberal markets, such as North America and Europe. A high percentage means that VoIP is strictly regulated and usually operating under a state monopoly.

North America

In the USA, regulating is handled by the FCC (Federal Communications Commission) and in Canada by the CRTC (Canadian Radio-television and Telecommunications Commission). Both institutions do not consider VoIP as traditional telephony and use a very liberal approach; no special licenses to operate VoIP are required.

However, as VoIP technology is growing rapidly, authorities are increasing regulations in parallel with traditional phone services. Here are a few of the requirements that should be supported by VoIP operators: number portability, 911 emergency services and permission to law enforcement agencies to access call information for court-approved electronic surveillance.

For up-to-date information, follow the news on the FCC and CRTC websites.

Latin America

Incumbents are dominating the telecommunications business here. However, local operators are not in a hurry to implement VoIP technology for two major reasons. First, circuit-switched networks are already established and generate a good profit. Second, when you already have a majority of the market in your hands, building completely new infrastructure to provide VoIP services suggests a questionable return on investment.

Europe

VoIP regulation in Europe is left to NRAs (National Regulatory Authorities). European Union (EU) countries are urged by the European Commission to take a "light touch" approach and develop their own treatment of VoIP. Requirements for VoIP providers are quite similar to those applied in the USA: support of emergency services, number portability, and provision of means to identify caller location if VoIP services are provided nomadically.

Asia-Pacific

The Asia-Pacific market consists of many contrasting countries with different penetration levels of VoIP, so each has its own approach. Regulation of VoIP is liberal and is allowed in most regions, such as Australia, South Korea, Thailand, and the Philippines.

Still, there are some exceptions. In Laos and Vietnam, VoIP is limited to specific providers, whereas in Cambodia no companies are currently allowed to provide VoIP (nevertheless, there's a chance that that situation will change soon).

Africa and the Middle East

Despite the high demand for cheaper international calls (the primary benefit of VoIP), the Africa and the Middle East (AME) region has the lowest VoIP adoption rate in the world. One reason for this is that telecommunications regulations have been established on behalf of telecom monopolies and consortiums which are trying to maintain their control of the fixed market. Another issue is that in some areas VoIP is technically not feasible because of a lack of requisite technologies, such as broadband Internet or Wi-Fi.

Usually, there is a direct relationship between a telecom monopoly and prohibition of the Internet telephony services. Because of this, VoIP is illegal or permitted only to licensed operators (monopoly incumbents) in many of the AME countries. Those who violate the law can be fined or even sentenced to jail. There are many cases where people have been arrested by police in Internet cafés for using or providing VoIP services illegally. There are just a few exceptions in this region where completely liberalized markets exist: Mauritius, Nigeria, and South Africa.

Even though liberalization of the AME market is slow, telecom monopoly operators are experiencing constant pressure to reduce prices due to international competition. Moreover, too-strict regulation and high calling rates are also reasons for illegal call termination (the routing of calls from one provider to another, or the connecting of calls to the end point or end-user) using VoIP GSM gateways (more information on this service in the "Wholesale Business" section of the next chapter). Such services are also called, "grey routes" or "grey termination." The concept of "grey" comes simply from the mix of the colors white and black, where white means that a call is originated legally, and black means that the termination is illegal. Because of grey routes, incumbents lose their voice traffic, even though they are constantly fighting against this by blocking SIM cards. Looking to the future, it is predicted that a grey market will continue to exist unless calling rates drop to the level of the USA or Europe.

Company Registration

VoIP providers usually register the company in the country where they do business (e.g., where the majority of their clients are located or where their employees work). Local business is supplied by all those providers that operate in the retail market (Internet service providers or cable operators). The situation is a bit different with international service providers. Theoretically they can freely choose the company registration location in a region with the lowest tax rate. But practically, that is not always possible, because it is in each country's interest that its citizens and companies all pay taxes locally. To avoid unwanted tax implications in the future, consulting a lawyer is recommended.

Starting a Company

Before starting a company, collect the following information from local business authorities and regulatory bodies:

- Is a special license required to provide VoIP services?
- What are the licensing fees and structure—annual, quarterly, or monthly?
- Are there additional taxes or fees involved in providing VoIP services?
- What reports should you provide for regulation authorities and how often?

If you decide to do business in one of the EU countries, there is one other thing to mention—the control of interconnection rates. The long-term vision behind this is to increase communication between people who are traveling within the European Union. Such a policy is end-user oriented. But from the business side, it decreases profit margins for operators and makes competition very tough for wholesale carriers that concentrate on EU destinations. As a result, many providers move their businesses to more profitable markets (usually in countries with telecom monopolies and grey termination).

Offshore Companies

There is a trend among wholesalers from heavily taxed countries (especially the CIS (Commonwealth of Independent States) region) to open offshore companies in countries such as Panama, the British Virgin Islands, the Bahamas, or the Seychelles, using offshore bank accounts. This is probably the most cost-effective way to operate a business. At the same time it can be risky as such activity is often against the law in that country. Moreover, some clients may not be willing to transfer funds to offshore accounts because of their local laws and restrictions.

Technological Requirements

Reviewing technological requirements in your market is a must before starting a business. There are many aspects to consider: location, certification, redundancy of the system, various standards, and support of number portability.

Location

Some regulatory bodies permit the use of on-premises infrastructure only. In these cases, it is not possible to take a chance on cloud-based solutions. This requirement is applied to keep more control over business operations.

Certification

Certifications are especially common in CIS countries. In this way, regulatory bodies are making entry into the telecom market harder and, effectively, are usually eliminating foreign competition for VoIP infrastructure. However, only bigger operators may follow this law, while smaller VoIP providers usually find some smart scheme to overcome it. One of the common ways is deploying two systems—one certified (shown to the authorities), another not certified, used in the real business. Such schemes are implemented because the certified equipment is either expensive or is not good enough compared to uncertified equipment. System certification is a long and expensive process. It results in the high price of certified systems as compared to uncertified ones, even though the only difference can be a certification sign on the equipment and stamped paper with a proof of this.

Redundancy

A set of rules that defines a minimal implementation which should remain operational even when the main system is down. This requirement is applied to minimize the risk of service failure. Even though implementing redundancy increases the investment to build a solid infrastructure, in the long term, it pays back.

Lawful Interception

The process of obtaining data for evidence or analysis by legal authorities. Information may be taken in real time or not (in case of the latter, it is referred to as "access to retained data"). In the European Union, call data must be stored for a specific period, ranging from six months to two years.

Quality of Service (QoS)

Refers to the overall performance (particularly that seen by the end-users) of a telephony network. Here are a few main points in discussions about the quality of service:

- What should be the minimal criteria for network quality and reliability?
- Should it be the same as the PSTN, or should it have specific requirements?
- What should be done with the information security to ensure users' protection and data privacy?
- Are there any special conditions for users with disabilities?
- This topic also includes password strength, encryption, and vulnerability testing.

Emergency Services

Each market has its own regulations of emergency services. In some cases, emergency service processes can be extremely complex and hard to implement. Here is an example of regulation in the Netherlands: if a user has a geographic number assigned to them, the service provider should check that user's postal code in the database and route the call to a specific emergency service provider assigned to this postal code.

Billing Process

The most important aspects, in this case, are billing time and invoices.

The billing time should meet certain standards. Usually, there is an organization or institute responsible for calibration and metrology (for the standardization of measurement) in a country or region. Some regulatory authorities require following a specific procedure to make sure that a measuring system meets their standard. so that 1 second measured is always equal to 1 actual second, and not 1.01 seconds.

Invoices should meet the minimal standards issued by a

tax agency. As a result, you need to choose a billing or accounting system that has the capability to issue invoices within such standards. An alternative option is to prepare invoices manually or by using third party tools.

Number Portability

Number portability refers to the ability to retain the same number if a customer changes service providers. There are two main types of number portability—Local Number Portability (LNP) for landline users and Mobile Number Portability (MNP) for mobile phone users.

Portability laws are administrated by regulatory authorities and are applicable in the majority of developed countries. It is important for you to get answers to the following questions:

- What is the procedure if a user wants to migrate to your network or from your network?
- Is this procedure integrated into your system?
- What is the institution that controls number portability?

Number portability is also important if you are interconnected with a local supplier which applies different rates depending on where calls were terminated—within its network or outside it. In this case, you need to control your routing based on a centralized number portability database, which relates a specific number to a particular operator to whom that number belongs at a given time.

Interconnections

Interconnection possibilities can differ depending on what type of license you have, and the restrictions applied to it. Some countries apply completely different regulations if there is a PSTN interconnection, so you need to keep this in mind and analyze it closely.

Chapter 2.
Choosing a VoIP Service

This chapter details what services exist in the VoIP market, how they differ from each other, and what is special about them. Note that there is no international standard for categorizing VoIP services and business models, so definitions in this book may differ from those you find in other VoIP-related information sources or that you hear from other people working in the VoIP industry.

Some of the services described in this chapter existed even before packet-switched networks, so you may wonder why those business models belong in the VoIP category. The truth is that in the majority of cases, VoIP business is not "purely" VoIP, because, on the path of a single call, it may pass through many different mobile, fixed telephony, and IP networks.

In general, there are two major categories of VoIP business—retail and wholesale. Most probably you have already used, or at least heard about, retail providers like Skype or Vonage that have become popular because their services are oriented to a mass audience. On the other hand, wholesale providers are rarely known because they operate in niche markets and usually directly serve other telecom operators or companies that generate more calls. The core differences between retail and wholesale lie in the target audience and quantity of calls. Wholesalers usually work in the B2B (Business-To-Business) market and serve those clients that can guarantee big call volume (telecom operators, retail providers, and bigger enterprises). Retailers can operate both in B2B and B2C (Business-To-Consumer) segments and deal with customers that produce lower call volume: residential clients, SOHO (Small Office and Home Office), and SMB (Small and Medium Business). However, there is no fixed breaking point between where a service provider becomes a wholesaler or retailer.

Providers

There are many different service providers: fixed and mobile operators, ILECs, CLECs, traffic aggregators and transit providers, calling card companies and call shops. However, all of them can be categorized into the three main groups: tier-1, 2, and 3. This section reviews the tiers, the role of voice traffic in retail and wholesale business models, and the purpose of telephone numbers, with a few use case examples.

Types of Providers

Tier-1 providers are big national and international operators. They are leading telecoms (Telefonica, BT, KPN, Orange, or Belgacom) which can ensure the most reliable network, the highest speed of data transmission, and the best quality of voice. Because these have the biggest network of telephone subscribers, the lower-tier providers seek to interconnect with tier-1 providers to have access to their network and to ensure the best voice quality. Even though the demand for interconnection is big, tier-1 operators are not interested in collaboration with small VoIP companies, because they do not meet the minimal order amount or have minimal traffic consumption.

This problem is resolved by tier-2 providers which collect traffic from smaller companies (ITSPs, calling card companies, call shop providers, or VoIP traffic traders) and terminate it through direct connections with tier-1 operators. Dealing with operators requires dedicated personal assistance, the highest priority of professionalism. In wholesale, this is usually handled by carrier relations managers (also called "account managers") who have the necessary skills.

Moving to the middle tier, quality of voice and data decreases. On the other hand, tier-3 providers are more flexible and can easily adapt to the changing needs of their market. Usually, they serve a local or niche target audience, which has specific demands that cannot be meet by tier-1 and 2 providers.

Even though there are only three levels described here, in reality the chain of providers can be much longer due to the participation of intermediate VoIP transit operators. There can be two types of providers involved in the chain—those that have their infrastructure and those that do not. Those that do are normally named as providers, carriers, or operators, and those that do not are resellers (also called "switchless resellers" to emphasize the fact they do not have their own switching system). (There is more information about the difference between resellers and providers in the "Reseller vs. Provider" section of the next chapter).

Voice Traffic

VoIP traffic (or "voice traffic," to speak generally about all

networks) is the flow of voice over an IP network that enables access to the PSTN. It is the main service supplied by all
VoIP providers. Successfully billed calls can be treated as a
commodity; the more calls that providers bill, the more profit they receive. However, in the real world, carriers try to
differentiate from each other by offering additional value-
added services and applying different pricing schemes, which
makes the final service more complex for the target audience.
VoIP traffic has a different meaning in retail and wholesale
business models.

Retail

The B2C market perceives VoIP as the technology that allows
free on-net communication and cheap international long-
distance calls. End-users do not use the term "voice traffic,"
preferring to think about it as a simple calling service. It is a
cost-driven market, making the pricing of VoIP calling service the key factor.

Rates can be based on a flat rate (also referred to as a
"flat-fee") or a regular fee. A flat rate refers to the fixed price
for a service, e.g. $40/month for the package "Unlimited
Monthly Calls to the USA and Canada." A regular rate depends on the usage, e.g. 4 cents/minute for a call to mobile in
Ireland. Each of the tariff plans can be provided in prepaid or
postpaid. Prepaid, or "pay-as-you-go," is a type of service
that does not require an agreement between client and supplier, allows the user to pay in advance for talk time, and top
up balance as needed (topping up is also called "increasing
the airtime" or "adding minutes"). Postpaid refers to the service with a contract between the customer and provider, with
a specific commitment and a defined period; billing is usually
done on a monthly basis.

A service which handles incoming and outgoing voice traffic in the B2B market is called "SIP trunking." It is provided to companies that have at least a few phones in their office that are connected to a communication system: a legacy PBX, an IP PBX, or a virtual PBX. SIP trunking will be reviewed in more detail in the "Corporate Services" section of this chapter.

Wholesale

The wholesale market has many definitions of voice traffic: call termination, routes, minutes or voice traffic transit. In general, it represents the call connection to a specific destination at a certain price and quality ratio. (More about this in the "Wholesale Business" section of this chapter.)

Telephone Numbers

There are many different types of numbers: geographic, local, national, nomadic, premium, toll-free, and virtual. Usually, the number range is provided by the national regulation authority for those operators that have a specific telecom license. Companies that do not have such a license may acquire numbers from providers, resellers, or directly from the licensed operator.

The best-known format that defines international telephone numbers is E.164. Such numbers are limited to a maximum of 15 digits. E.164 numbers usually start with the plus sign (+), followed by the international country calling code. A number such as 020 7100 1111 in the UK would be formatted as +44 20 7100 1111.

Most VoIP platforms and devices support E.164 to allow

calls to be routed between VoIP and PSTN networks. Also, VoIP implementation may use additional identification techniques, such as SIP-URI, used as a SIP phone number that is similar to an email address. It is displayed in the following format: sip:x@y:Port, where x is a username and y is a host (can be an IP address or a domain name). For example, sip:joe.smith@123.123.1.10, sip:sales@pbx.ip.com, or sip:123 456@pbx.ip.com:6000. Such a format can be translated to the E.164 and vice versa.

Private telephone networks may have their own numbering plans. Such private numbers are also called an "extension," usually a short three- or four-digit number. The internal communications between telephone extensions are controlled by a PBX (private branch exchange). Calls between extensions are free, allowing greater savings, especially if a company has branches in different cities or even countries. Moreover, a call connection between extensions is very fast since all communication is done inside the private network.

Probably the most common these days are virtual numbers. They are also named "direct inward dialing" (DID) or "direct dial-in" (DDI). DIDs can be used as an access number, for incoming calls or call forwarding.

DIDs are widely used in a majority of retail businesses and are the core component in services like a callback, calling cards, and call-through. In this case, DID acts as a gateway between the PSTN and a VoIP network. (More information about services that use DIDs in the next section.)

Another use of DIDs is as a possibility to receive calls. For example, take an organization with 100 employees, where each of them can have a phone, reachable by an extension. The probability that all 100 employees will be called at the same time is very low. In this case, the company could take one DID number with ten channels (also called "lines"), which allows ten concurrent calls (CC). If all channels are in use, a caller will have to wait for an available line, will get a busy signal, or will be forwarded to voicemail. Such a scenario increases the communication efficiency and allows reduced costs on phone lines.

Finally, DIDs can be used for call forwarding to an external number, VoIP phone, ATA adaptor, IP PBX, VoIP application on a smartphone or computer, voicemail, or fax. A virtual number is not directly associated with a telephone line and can be used in any geographical location. Because of this, DIDs are very popular among migrants who want their friends and family to reach them by a local phone number, and those who want to have an affordable global presence.

The last term that is widely used when talking about telephone numbers is a Caller ID (in short, CID). It is the number that the call recipient sees on their phone screen when someone is calling him or her. Most providers transmit the original Caller ID of the caller, but this number can be easily changed.

End-user Services

This section reviews VoIP services for the following segments: home users, mobile subscribers, and public place visitors.

Some of the retail VoIP providers promote their services by using the following names: Device2Phone, PC2Phone, PC2PC and Mobile2Phone. Those are definitions of services that describe the method of call initiation and termination:

- **Device2Phone**: calls are initiated from a hardware device like a legacy phone or an IP phone, and terminated in the PSTN.
- **PC2Phone**: calls are initiated from a softphone (a software program for making telephone calls over the Internet) using VoIP, and terminated in the PSTN.
- **PC2PC**: calls are between two softphones within the same VoIP network.
- **Mobile2Phone**: calls are initiated from a mobile dialer (a mobile application for making telephone calls over the Internet) using VoIP and terminated in the PSTN.

Residential VoIP

Residential VoIP is oriented to those who are calling from a fixed location: home users, home offices, and the self-employed. They can initiate calls from both hardware devices and software applications. In this subsection you'll learn more about each of the calling methods and services which come as a package together with VoIP calling.

Triple-play

VoIP technology has enabled small Internet service providers, cable, and fixed-line companies, to compete directly with the local incumbents. Small players have the opportunity to expand their service portfolio by offering a combined package, consisting of Internet and voice services. The optimal solution for providers that deal with the residential market is to offer the "triple-play," which means provisioning the combination of three "bundled" services: broadband Internet, television, and VoIP services.

This combination of different components allows providers to upsell VoIP easily by mixing it together with other services. In very competitive markets, ISPs offer VoIP services at cost (or even go into some loss) in order to encourage their clients to sign up for the Internet service. In this case, VoIP becomes a tool to support their core business—Internet connectivity.

Analog Telephony Adapters (ATAs)

Despite the popularity of VoIP technology, there are still many users who have a traditional phone (also known as a "landline," "legacy phone," or "fixed-line"). To start using VoIP services, it must be plugged to the ATA. It usually has two types of ports: one to connect to the Internet (DSL or cable modem) and another for connecting analog telephones. Here you can see a picture of phone adapters that are very convenient for home or small office usage.

Figure 2.01. Phone adapters.

ATA allows the use of traditional phones in the same way as VoIP phones. Usually, such adapters are preferred by those users who are reluctant to change their calling behavior and do not want to spend more money on VoIP phones (ATAs are usually less expensive). Some telephony service providers, such as MagicJack or Vonage, provide branded ATAs to their clients as part of the service package to make unlimited phone calls to the United States and Canada.

Another device similar to an ATA is a VoIP PSTN gateway. Compared to an ATA, gateways usually have more ports and are used by corporate users and service providers. A few examples will be reviewed in the next section, "Corporate Services."

VoIP Phone

Those who are looking for more advanced features can choose an IP phone. It is a device which uses packet-switched technology for making and receiving calls. It looks much like an ordinary telephone but is used specifically for transmitting telephone calls over an IP network and has a richer feature set. There are many types of phones: corded, conference phones, video phones, Skype phones, and USB phones. Here you can see a picture of an IP phone.

Figure 2.02. IP phone.

However, the majority of home users do not need fancy features that come, usually, at a higher price. Because of this, IP phones are not as popular in residential VoIP and are mostly used by the corporate segment.

Softphone

Softphone (referring to a software phone) is an application which runs on a personal computer, for initiating and receiving calls. It is usually free, or at least a more cost-effective alternative to the hardware phones. Softphones are technically called "SIP user agents" or "SIP clients." Here you can see a screenshot of the Zoiper desktop softphone.

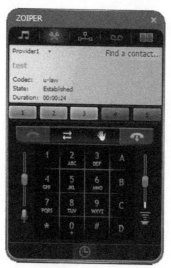

Figure 2.03. Zoiper desktop softphone.

Softphones are usually associated with home users, but are also a good choice for offices which require a convenient replacement for traditional desk phones.

SIP Account

To start using a VoIP service, a customer needs a SIP account which is generated by their VoIP provider. A SIP account establishes the specific settings that need to be entered to a SIP-compatible device or application to use VoIP services. The usual settings to register the device and authenticate the customer are:

- SIP proxy (it can be also referred to as "domain name" or "IP address." It refers to the address of the SIP server or a softswitch which authenticates the customer)
- Username and password (credentials used to register the device)
- Protocol and codec (the most common are SIP protocol and the G.729 or the G.711 codec)

There is a huge market trend for smartphones and more advanced users prefer to be mobile and have their main tool of communication with them wherever they are. Because of this, mobile apps have become a more popular alternative to softphones.

Mobile VoIP

Mobile subscribers are the biggest audience due to the high level of mobile device penetration worldwide. This segment includes everyone who calls using mobile phones. Mobile users can be divided into two categories according to the devices they use. Simple mobile phone users stick with an "old-fashioned" way of calling by dialing an access number. Smartphone users use mobile apps for calling purposes.

Call-through and Pinless Dialing

Call-through is done by dialing the destination through an access number. The caller can be authorized by the Caller ID or PIN code. The first method is also called an "automatic number identification" (ANI) and is usually the method most preferred by clients because it is faster and simpler. Sometimes it is called "pinless dialing," emphasizing that the user does not need to enter the PIN code. The second method requires entering the PIN code before dialing the destination. It is more complicated for users since it prolongs the calling process. However, it ensures a higher security level because otherwise the Caller ID can be faked by using a Caller ID spoofing method.

The call flow using an access number looks like this:

- The caller dials an access number.
- The mobile operator establishes a connection with an access number.
- The access number (DID) provider automatically forwards calls, which are passed from the DID to the retail VoIP provider to which this number belongs.
- The retail VoIP provider automatically forwards this call to a wholesale VoIP provider.
- The wholesale VoIP provider makes the final connection with a call recipient.

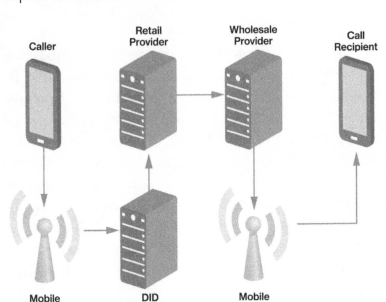

Figure 2.04. Call flow using an access number (DID).

The caller pays for two suppliers: the retail VoIP provider for making a long-distance call, and the mobile operator for calling to an access number. Calls on the same network are usually free or charged at a lower fee, so retail VoIP providers must make sure that access numbers of different mobile and fixed networks are available. (Such numbers are usually acquired from a company (a DID number provider) that specializes in DID aggregation.) Mobile access numbers are dedicated for those who call from cell phones, while landlines are reserved for those who call from landlines and payphones (phones in public places). An alternative method is using a toll-free number, which is free for both mobiles and landlines. However, toll-free numbers are usually more expensive than other DID numbers, which means that either the calling rate to the final destination will be higher, or there will be some fee for calling to a DID number.

Dialing an access number before a call can make the process more complicated than a regular call. So what can be done to make it more convenient for the customer? Here are some shortcuts that can make the life of your clients easier.

First, the number can be saved in a special format in a phone book (access number at the beginning, followed by a pause and then an international number). Here's an example of the German access number to call to the UK destination: 0049123456789P0044123456789. Such procedure can be done on any brand of mobile phone.

A second solution is using a SIM dialer. It is an attachment to a SIM card that can store a few access numbers. As a result, the user's conventional dialing habit can be retained to simplify the calling process. A user can dial directly from the phone book without needing to input access numbers as suggested in the previous example. SIM dialers are compatible with a majority of mobile phones in the market.

The third option is to use a mobile app (mobile dialer) which stores the DID and dials it automatically if the user makes an international call.

Mobile Dialer

The mobile dialer is an application for smartphones that allows the sending and receiving of VoIP calls. You may be familiar with the strongest and most well-established brands offering VoIP through a mobile app, such as Viber, Skype, Google Hangouts (Google Voice), MagicApp (by magicJack), and Vonage Mobile. They are also called "over-the-top" (OTT) applications. OTT refers to the service that is used over the network of the provider (e.g. mobile operator).

Each mobile dialer has its strengths and weaknesses, but

their main purpose is the same. There is a "buddy" list of contacts to whom you can send instant messages (IM) or place voice calls over mobile data or Wi-Fi for free. Since these applications are usually free, their creators apply extra fees for the premium features. The most common is a per-minute rate or a flat fee for calling outside the network to landlines or other mobile phones. In addition to this you may obtain a permanent phone number, get an extended conferencing capability, voicemail, or securely encrypted calling features for a small monthly subscription fee. Here you can see a screenshot of the Zoiper mobile dialer for Android.

Figure 2.05. Zoiper mobile dialer for Android.

Calling from the mobile application using VoIP technology has become more popular with the introduction of lower-cost mobile data and the mass deployment of GPRS, 3G, 4G and Wi-Fi networks. You now can run a dialer on all mobile platforms: Android, iOS, Symbian, Windows Mobile, and BlackBerry. Calling is made very easy; it can be done directly from the phonebook.

As VoIP is mostly used for international calling, it is possible to make a configuration which allows the use of a mobile dialer as the default application whenever you call an international number. Moreover, if a phone is in a zone with no Internet and mobile data is very expensive, there is the possibility of calling from the mobile dialer by using a DID number. Unlike callback and call-through services, DIDs are stored inside the application, and the calling process remains the same. Such capabilities make mobile dialer a single point of communication; it unifies services like SIP-to-SIP, SIP-to-PSTN, call-though, and callback.

NOTE FROM THE AUTHOR: Nowadays people can choose to message or call someone using OTT applications in a smartphone rather than paying their provider for calls and SMSes. It is forecasted that the number of mobile users will reach 6 billion by 2020, and the number of Internet users will reach 4.7 billion. People will access the Internet primarily through smartphones which will become, ultimately, the default mobile device.

This trend affects services like calling cards, call shops and fixed-line. Those markets will decrease as people prefer to be mobile, global and flexible both in their lives and at work. However, mobile apps supplement traditional voice and SMS services rather than replace them completely. I'm sure that the PSTN market will still have a significant market share for quite a long time.

Callback

Callback occurs when the caller is called back after initiating a call through an access number. Callback is used in two main cases: to deliver services to those countries where DIDs are not available and to provide services for those clients who prefer not to pay for the call to the DID.

In many countries where the telecom market is strictly regulated, telecommunications services belong to a single company controlled by the government. In such environments, other providers are not allowed to offer their services and it is not possible to obtain a local DID number for access number purposes. In this case, callback is the only choice that allows local people to save lots of money on international calls. In other regions, where DIDs are available, callback is rarely used unless there's an economic reason (DIDs are expensive or calling rates to DIDs are high).

The complete process of a callback service usually looks like this:

- The caller dials a DID number.
- Call hangs up automatically and the caller hears a busy signal.
- Within a few seconds, the caller receives a return call.
- The caller picks up this call and hears a message prompting them to enter the destination number.
- Once a destination number is entered, the connection between calling parties is established.

The cost of making a telephone call via callback consists of two parts; the caller is effectively paying for an outbound and inbound call at the same time. For example, a customer from Senegal is calling Venezuela. The local mobile operator would charge $2/min for a call to Venezuela, but as a connection was not established, there's no fee. With callback service, the provider charges a customer for two calls:

- Inbound call to Senegal ($0.3/minute), also called "Leg A." This is the call returned to the subscriber after the trigger call is initiated.
- Outbound call to Venezuela ($0.1/minute), also called "Leg B." This is the connection to the destination number.

The customer pays a total of $0.4/minute for a call from Senegal to Venezuela. Compared with rates of a mobile operator, the savings amount to 80%.

Calling Cards

The essence of the calling card business model is to creatively segment customers by various demographics or behavioral characteristics and to design calling card offerings to meet specific calling needs. This business typically attracts entrepreneurs who want to enter the new market, businesses with established retail distribution channels, and service providers that want to diversify their revenue streams.

A calling card is a small plastic or paper card, sized and shaped similarly to a credit card, used for making low-cost long-distance calls. On the front of the card are the brand name and the card price. Usually, the design depends on the target audience for which the card is dedicated.

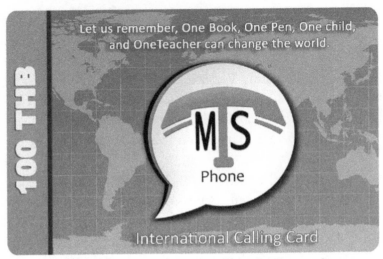

Figure 2.06. Front of the "MTS Phone" calling card.

On the back of the card there is usually the following information: an access number (or a few of them), PIN (it is hidden by a scratch panel and must be entered to activate the card), calling instructions, terms, conditions and a customer service number.

Figure 2.07. Back of the "MTS Phone" calling card.

Here's an example of a call flow using a calling card:

- The customer dials an access number.
- The customer listens to a message prompting them to enter a PIN—or, if the card is pinless, the message will ask for the destination number.
- Once the PIN is entered, a system announces the account balance to the caller and invites the caller to dial the destination number.
- Once the destination is entered, a system announces the amount of minutes available for that destination.
- The customer is connected with the destination.

Calling cards are usually sold in any built-up area frequented by residents traveling overseas, ethnic groups, tourists, foreign students and military personnel. The success of this business depends and fully relies on distributors. They are responsible for delivering cards to retail stores, local shops, post offices, petrol stations, kiosks, or selling them over the Internet. Currently, the rate of calling card usage is decreasing, so it is highly important to make them available and handy to buy.

By using calling cards, people try to save money on their long-distance phone bill. In fact, there are no monthly bills; customers simply pay for the calling time in advance and the balance of a prepaid calling card gets reduced after each call. Calling cards are usually chosen based on calling needs, such as countries and call lengths. However, exact calling rates are usually very tricky. Cards can be divided into "white" (clearly stated fees) and "black" which may include various additional fees like:

- Connection fee
- Disconnection fee
- Multiple in-progress fees
- Post-call surcharges
- Maintenance charges
- Custom rounding increments

All of those fees, if applied well, maximize the calling card provider's profit. However, not all potential customers respect a provider that adds many hidden fees; they may prefer to choose a more transparent provider. Nevertheless, a majority of customers are still attracted by a low price and will then be persuaded to buy "black" calling cards, whatever the fees. Providers should do a lot of statistical analysis and testing with the target audience to find out the most profitable way to produce calling cards.

Call Shop

A call shop is a place that provides on-site access to telephones for long-distance calling. The "real" call shop is meant only for calling. However, there is a low probability that you will find such a call shop these days because the rent and the salary for a call shop manager would be too expensive to justify running a business that only enabled people to make long-distance calls.

To support business expenses, the service portfolio should be wider, and calling services are a secondary source of income for places such as Internet cafés. Such places are usually dedicated to an ethnic community, immigrants, and travelers who want to use Internet services, make long-distance calls, or do a money transfer to their home country.

A call shop requires the following:

- Calling booths
- Cashier's desk
- Calling equipment
- Billing software
- Internet provider
- VoIP provider

Because all calls are personal, the call shop owner must ensure private and comfortable spaces for calling booths. Each booth should contain calling equipment (IP phone or traditional phone, connected to an ATA adapter). Moreover, it is recommended to have a display where customers can keep track of call duration and costs. This gives more confidence to a client to control what they spend.

A cashier's desk usually requires a computer to access the billing interface (to control calls in booths and generate bills) and a printer for invoices.

Finally, a call shop should have access to the Internet and the possibility to make international calls. For that, an agreement with an Internet supplier and VoIP provider is necessary.

Corporate Services

Business clients include all small, mid-size, and large companies, institutions, and organizations. They bring more income than residential users, but the higher income comes at a higher risk. If a large corporate client terminates their contract, it will have a huge impact on the service provider's revenue. Because of this, such clients are treated on a higher level than others. Service providers try to increase the loyalty of corporate clients by offering a wide range of services: voice traffic, Internet connectivity, telecommunications devices and infrastructure, virtual numbers, and hosted PBX.

There are also many niche businesses that require tailored solutions, like call center management, call conferencing, or unified communication platform. Such clients tend to choose a supplier that can offer a bundled package. Having expertise in a specific business segment (call centers, real-estate agencies, or government institutions) allows a service provider to adapt and fine-tune a complete package for clients. In this section, we'll go through the main services that are provided for business clients.

IP PBX

Most companies that have at least a few phones use a PBX. It is a telephone system meant to manage and control internal communication within the company and external calls to the PSTN. PBX acts as a gateway between company and service provider and enables such functions as:

- Local extensions to reach individual employee
- Free calls between local extensions
- External phone calls to the PSTN
- One DID number of outside lines
- Value-added features (voicemail, ring groups, call transfer, call forwarding, conference call, and auto attendant)

There are different types of PBXs: hardware or software,

premise-based or hosted, traditional or IP-based, open source or proprietary.

Hardware vs. Software

Hardware-based PBXs come as physical equipment with pre-installed software. Software-based PBXs can be installed on any computer or server that meets the minimal hardware requirements.

The main difference is that hardware-based solutions are usually more expensive, but at the same time you get a complete solution from a single provider, which makes resolving problems much simpler. There are two types of issues: software-related or hardware-related. Software-related problems can be resolved remotely, whereas hardware-related issues can be fixed by physically replacing part of, or all of the equipment.

There is nothing to be concerned about if such a process is fast and there is no need to ship and wait for a replacement. But if a manufacturer is located overseas, it can cause a big delay and present a huge risk for the business. As a software solution can run on any computer, changing a component is less expensive and much faster compared to replacing a specific part of the hardware which may be produced by only one manufacturer. On the other hand, even though fixing hardware-based solutions is more expensive and takes more time, it bypasses situations where the software supplier blames the hardware supplier for causing problems and vice versa. Such a situation is confusing for the infrastructure user and does not leave any other option except to take control of the communication between two suppliers.

On-Premises vs. Hosted

An on-premises PBX resides at the customer's office, usually in a wiring closet or computer equipment room. Hosted PBX (also called "hosted VoIP," "cloud PBX," or "virtual PBX") is deployed on the service provider's premises, and customers usually access it through an online management portal. The current trend is to move infrastructure to the cloud, so virtual PBXs are becoming more and more attractive. However, companies still face a dilemma over whether to choose on-premises or cloud-based solutions.

The on-premises model is usually better for those companies that have a skilled technical team who can take care of infrastructure management and prefer to have more control of the system and confidentiality of business data. In such a case, implementing a new system locally and managing it does not become as big a challenge because such companies already know the possible hardware-related issues and how to overcome them.

On the other hand, those who have no infrastructure in place, no technical team, and no knowledge about server management, should consider cloud-based solutions. They allow companies to focus on their core business operations instead of managing the technical side of communication.

The table below outlines the pros and cons of the on-premises and cloud-based delivery model.

	On-premises	Cloud
Quality	It is recommended to have the PBX as close as possible to the	Cloud-based solutions are limited to a few locations, so

	caller. Using an on-premises PBX ensures the shortest calling path that results in the best call quality.	clients may not get a server close to their office. Long distances between caller and PBX can cause delays that have a negative impact on call quality.
Connection	If a customer uses a conventional PBX, the connection does not depend on the Internet. If the Internet is down, the client will still be able to make and receive calls.	Loss of the Internet results in loss of phone service. However, settings can be adjusted, and incoming calls may be forwarded to voicemail or a cell phone.
Costs	There are higher initial purchasing and set-up costs. System and power backups, feature upgrades, and maintenance usually come at an additional price. However, system ownership may pay back in the long run.	Based on a monthly subscription that usually includes upgrades, support, backup systems, and server maintenance. It is easy to switch to a lower or higher plan depending on the number of phone lines in the company.

Control	The client has physical access to the system and more control over management.	The system can be accessed only through the web panel, so the client has less control over management.
Confidentiality	Corporate data is stored and handled internally.	Data is accessible by the service provider as well.
Moving	If a company changes their office location, there will be downtime while the PBX is physically moved. Once the equipment is deployed, someone will need to take care of connecting phones to a PBX.	There is no need to move a PBX physically, so changing an office is easy and there is no downtime. Calls can be temporarily forwarded to the mobiles while moving IP phones.
Maintenance	Server maintenance is done internally by dedicated IT staff.	Server maintenance is handled by the service provider.

Legacy PBX vs. IP PBX

Legacy PBXs are based on the traditional telephone network while IP PBXs use IP networks to transmit calls. Over the last couple of years, the traditional BRI and PRI lines have started to become rapidly replaced by the VoIP service, the growth of which does not seem to be slowing down. This results in greater opportunities for service providers focused on delivering IP PBX (especially cloud-based) and SIP trunking services to SMBs and enterprises. By using SIP trunking, companies can replace the conventional telephone trunk. There are two ways to do that: connect the legacy PBX to a gateway that converts SIP to PSTN (and vice versa) or use an on-premises or cloud-based IP PBX.

As you are already familiar with the benefits of VoIP (they were listed in the "Voice over IP" section of the "Understanding Telephony" chapter), you may be interested to know the reason some companies still use legacy telephony systems. One of the telecom veterans, Richard "Zippy" Grigonis, shared the results of a Level 3 study in a TMCnet article, listing the top five reasons for not switching to VoIP:

- Satisfied with current service
- Was never offered a special incentive to switch
- Not aware of the cost of the service
- Not familiar with the provider
- Concerned about VoIP's quality and reliability

Those points show what scares business clients away from switching to VoIP. With this in mind, you could create a strategy and a list of questions for yourself before approaching such customers. Take your time to answer the following questions right now if you are planning to offer IP PBX for corporate clients:

- If they are satisfied with their current supplier, how will you create a need for them?
- Switching to another provider can be very painful.

What is in your service that would encourage them to make a change?

- If they are interested in cutting costs, how much will they save after they start using your services—in a month or a year? When will the investment pay off?
- Is your brand and name well known? If not, do you have references that would prove you are doing a great job?
- How can you ensure high voice quality? What will you do if their Internet connection goes down?

As you see, you should demonstrate credibility, reliability, and a high value of VoIP versus traditional landlines to gain the trust of a business client.

Open Source PBX vs. Proprietary PBX

The first telecommunications systems were only proprietary, with a closed source. The competition was not too heavy, resulting in high prices and few businesses which could afford it.

However, things changed once open source technologies appeared in the telecommunications sector. Prices dropped as companies were able to take a simple computer, install an open source platform such as Asterisk (an open source PBX), and turn it into a PBX. When deciding between an open source or a proprietary solution, consider a few important aspects: price, quality of code, customizations, maintenance and security.

Cost is the biggest advantage and one of the core reasons why people tend to choose open source solutions. Users have the ability to download the latest telephony features from community portals for free, making open source systems the most affordable and flexible option on the market. However, if you download software for free, it does not mean you will not experience any expenses associated with it. If you do not have sufficient technical knowledge or skilled personnel, services such as installation, configuration, customization, and maintenance can be provided only by professionals who charge for their work. Even though the proprietary software is paid, you will get a complete service range directly from one supplier that can also give you more guarantees.

The second aspect is the quality of code. Open source software is based on community customizations, and the code is written by various developers worldwide who are usually doing this work in their spare time. Because of this, the overall software is not well-tested, leaving you with the full responsibility of launching it into production. Moreover, those programmers are not getting paid for their work (which is natural since the software is free). This removes from you the right to blame someone else for a bad quality of code, or to hold them accountable.

In the end, there are two choices: either report a programming error (also called "software bug") to a community forum and wait till it is fixed or revised in a newer version or pay an experienced developer who can fix it faster.

Another option is to choose a commercial system. It does not mean that it will not include any programming errors, but at least there will be a company behind it which is responsible for fixing such issues fast, so your business does not experience negative effects. Usually, companies that build commercial software have highly efficient programming and testing processes and hire experienced developers who are being constantly trained to ensure the quality of the software.

Both open source and commercial software are continuously developed and from time to time, updates and new versions are released.

However, new features are oriented to global needs and may not meet the specific requirements of your business. If you miss some important function, you can request a customizing service. In the case of open source software, you need to choose a developer, a majority of whom are freelancers. Commercial software providers can add requested features by themselves. The advantage of freelancers is that their work costs less, they are more flexible, and their customizations are done faster.

At the same time, there is a bigger risk because your business depends on one person. Imagine what you would do if the freelancer you hired stopped responding to you? Keep in mind that finding another programmer who will review and fix someone else's work would be time-consuming and more expensive than ordering the primary feature implementation. Customization done by a commercial organization usually takes more time, but at least this option is more reliable and includes some guarantees and a support agreement.

If the proprietary software developer has a global product, there is a big chance that the custom features requested by individual clients will not even be considered. In this case, open source solutions become a much better alternative for those who have unique needs and whose vision does not match the vendor's roadmap.

Systems maintenance (or technical support) is similar to

customizing. There are few options to get technical support services: use a public forum, have a knowledgeable team, or hire a professional technician for direct assistance. Using a public forum is free and is a great option if the community is large enough; there will always be some active person with experience who can answer your questions and propose a fast solution. But imagine the situation if you have a major problem which causes huge losses for your company in a short time (software crash, some serious bug, or a hacker in your system). I do not think you could open a new thread in the forum and patiently communicate with other forum users. In this case, you need to resolve your problem as soon as possible, and the only way to do that is to prepare in advance. If you use an open source system requiring constant maintenance, consider dedicating a person in your organization for that purpose. If problems arise only from time to time, you could make an agreement with a few freelancers who will help you in case of a critical issue. (I'd recommend you consider enlisting a few people, in order to have "human redundancy," in case one person is on vacation, sick, or otherwise unable to respond promptly.)

On the other hand, if you use proprietary software, it usually comes with a certain service level agreement and a team of engineers who will be able to react fast. Moreover, commercial platforms are usually better documented, since this reduces work for a technical support team. For frequently asked questions (FAQ), they can often point clients to some video or documented instruction.

The last of the criteria is system security. Open source platforms are quite popular, and the code is visible to everyone. Because of this, they are vulnerable; the hacking community is familiar with security holes much better than a new user who is just starting to deploy an open source solution to a production environment. Commercial software is more protected and far less susceptible to hackers.

SIP Trunking

A trunk refers to a line or link that connects a company's PBX system with the service provider's softswitch by carrying many signals at once. As SIP is the most common protocol in IP telephony, the service that facilitates the connection to VoIP is usually referred to as "SIP trunking."

SIP Trunking vs. Legacy Telephony

There is a trend in most countries for legacy telephony to move from the PSTN to much more modern and flexible IP telephony. Over the last couple of years, SIP trunking has grown rapidly, replacing analog ISDN BRI and PRI lines, and it does not seem to be slowing down. Many of the major telcos have already stopped offering customers ISDN services or plan to do so in the near future. Moreover, the authorities in some European countries have even issued laws requiring that businesses must switch to VoIP within a few years. Taking all this into consideration, the telecom market research firm Infonetics predicts that the SIP trunking market should reach $8 billion by 2018. This opens huge opportunities for VoIP service providers to approach the corporate segment.

The problem that companies face with the traditional BRI and PRI lines is mainly associated with costs and flexibility. If a business reaches the limits of one PRI line, there is a need to add another one with additional 23 channels (according to the US standards), which is very expensive and inflexible. This can easily be solved by using an alternative solution—SIP trunking, which has some key advantages. Companies receive lower calling rates (usually to international destinations) and reduce expenditure on phone lines. BRI and PRI lines are much more expensive than SIP channels which can be easy scaled by allocating more bandwidth. Moreover, using SIP does not require purchasing a "package" of lines, as in legacy telephony. Service providers can offer adding channels one by one.

Provider's PBX vs. BYOD (Bring Your Own Device)
To gain access to a VoIP service, a company must have a PBX and sign an agreement with a SIP trunking provider. It is important to note that VoIP providers may have different approaches, according to the corporate clients they deal with. The main ones are:

- Virtual PBX
- On-site PBX
- BYOD Policy

Virtual PBX is the best option for the SIP trunk provider as it gives the most control on call quality, and all issues can be resolved remotely. Moreover, it makes the SIP trunk provider a "one stop shop" for all services including voice traffic, DIDs, and virtual telephone exchange. This makes the client's life easier since all communication products and services are provided by a single supplier, allowing the provider to increase the revenue stream by combining everything into one package.

Another option is to deploy an on-site PBX. To perform installation and configuration tasks, VoIP providers usually partner with system integrators who are responsible for organizing all the technical work and PBX maintenance. In this case, it may take more time to troubleshoot problems, as not all issues can be resolved remotely. The more physical maintenance that is done, the higher the self-cost for a service provider or a client (depending on the warranty period and type of issue).

The last option is the BYOD policy. This means the company can use its own PBX with the provider's service. This policy usually opens doors to more potential clients as most companies have already invested in their communication infrastructure and are eager to retain it. However, the BYOD policy brings more responsibility, especially on call quality. The provider's softswitch should be compatible with the client's PBX. Moreover, if a company uses legacy PBX, there is a need for a gateway that will convert PSTN to VoIP and vice versa.

VoIP Gateway and Telephony Card

If a company has a legacy PBX and wants to subscribe to the SIP trunking service, they will need to connect the PBX to a VoIP PSTN gateway. A gateway (see picture below) provides the interface for connecting different networks and allows a smooth transition between the PSTN and VoIP.

Figure 2.08. Yeastar PRI VoIP gateway TE100.

An alternative solution to the VoIP gateway is a BRI/PRI card (also called a "telephony interface card" or, in short, a "telephony card"). However, this can be used only for software-based telephony systems that run on a server or computer. Because of this, such cards are used for a different purpose – to connect open source communication platforms (like Asterisk, FreePBX, or FreeSwitch) to E1, T1 or ISDN-BRI interfaces.

Here you can see a picture of a telephony card in a data-center.

Figure 2.09. Telephony card.

Even though telephony cards are cheaper than gateways, they may require additional drivers to be loaded into the machine's operating system. This may involve additional manual work, requiring some technical knowledge. For a more convenient setup, it would be better to connect a simple gateway.

Unified Communications

Recently, telephony solutions aimed at SMBs have evolved into unified communications (UC) services. UC is the integration of multiple real-time communication services, such as voice, video, instant messaging, conferencing (audio, video, and web), desktop sharing, and data sharing. It is meant to provide a unified user-friendly interface for communication across multiple devices and media types. As a result, it optimizes business processes and increases user productivity.

People use many different devices and applications to communicate. Moreover, communication comes in different forms. Enterprises organize video conferences, which are a more cost-effective and efficient alternative to personal meetings. Sales managers make online presentations for potential clients through desktop sharing tools. Finally, most people use email and instant messaging applications to chat more often than they use voice communication. I am personally in touch with clients from 90 different countries, each of them having a separate English accent. Exchanging emails or text-chatting through Skype often allows much better understanding.

Now think about all those devices (mobile phones, IP phones, personal computers, and laptops) and applications that are used in the business environment. Multiply them by all the media types and you'll get a pretty big matrix of choices. Some businesses can live with that. But when the diversity of communication processes grows to a certain level within the enterprise, it becomes a pain. This pain is solved by choosing the right unified communication platform that can be adapted to the needs of the enterprise. The ultimate goal is to break down barriers so employees can communicate with anyone, anywhere at any time. Unified communications (UC) encompasses a collection of different elements including (but not limited to) those described below.

Call Control

Call control functions are usually handled by the IP PBX; in the bigger picture, this comes as one of many elements in unified communications. It allows the routing of calls to the proper recipient efficiently, to transfer calls if needed, to review the calling statistics and to perform other functions related to calls.

Unified Messaging

Unified messaging is the integration of different electronic messaging media: email, SMS, fax, video messaging, and voicemail. All types of messages are stored in one system and can be accessed from a single interface.

Instant Messaging

A type of online chat that offers real-time text communication over the Internet or a private connection (if a company uses it only for internal communication). More advanced instant messaging incorporates extra features like voice over IP, video calls, transfer of files, or clickable hyperlinks.

Presence

Presence is the real-time indication of a user's availability. Most people associate presence with the actual period of instant messaging. However, it can also be used in business process applications to find out if the right person is available at a certain time.

Conferencing

There are three types of conferencing: audio, video, and web. Audio conferencing is used to add multiple participants to a call. Video conferencing provides all the advantages of a face-to-face meeting, all the while cutting travel costs and improving productivity. Web conferencing is usually used for webinars, product demonstrations, training or technical support.

Collaboration Tools

Collaboration tools include applications like workflow, calendar, and schedules. All of them help individuals and workgroups to collaborate with colleagues, customers, or partners, via the Internet in real time. Collaboration tools are widely used in sales and customer support and allow the sharing of desktops or particular websites while offering voice, video, and text chat assistance.

Mobility

Mobility allows access to real-time communication services from mobile wireless devices, regardless of the user's location. It is especially useful for companies that have remote workers, giving them similar capabilities as if they were in the virtual office, thus improving their availability and productivity.

Business Process Integration

Each company has many different business processes and workflows. Projects are often delayed due to "human latency," which is not easy to track and control. However, if managers can access the complete project through a single interface and see who is working on the project at a particular time, it becomes much easier to manage and improve the overall process. UC is becoming a part of business process applications such as ERP (Enterprise Resource Planning), supply-chain management, CRM (Customer Relationship Management), and sales force automation.

Another trending standard related to unified communications is WebRTC (Web Real-Time Communication). It acts as a communication gateway for any browser and enables a wide range of services, including calling, conferencing, instant messaging, and video streaming. WebRTC does not need any plugins, so it will eventually replace web phones (applications that allow users to call directly from a web browser). Even though it is still a new concept for many enterprises, it is expected to grow rapidly.

Wholesale Business

The wholesale VoIP market is very competitive and offers low profit margins. To have a sustainable business, high voice traffic volume must be managed. There were good times for wholesale. I recall speaking with one of my clients from Russia, who had bought an apartment in Moscow after three years of operating in the wholesale VoIP business. He joked back then that he had already earned his pension, so now it was time for him to work for something else. Now it is not that easy to achieve such significant growth in wholesale, but there are always new opportunities for those who keep their finger on the pulse of the VoIP market.

Before digging deeper into the specifics of wholesale, it is important to understand the role of participants in a wholesale VoIP chain. There are three main types of services: origination, transit, and termination. A call originator is a retail VoIP provider that initiates a call (usually PSTN to VoIP). A transit operator is a company that passes the call from the originator to terminator or another transit provider. A call terminator is an operator that makes the final connection with the call recipient in its network or the PSTN.

This next section describes the participants of the wholesale chain and different wholesale business models.

Origination

Call origination refers to the collecting of calls initiated by end-users. They can either call from VoIP devices connected to PBXs or from mobile phones using access numbers. This type of voice traffic is usually originated by retail VoIP operators or DID providers. All services that were described in the "End-user Services" and "Corporate Services" sections can be referred to as "call origination." Also, it is important to note that the term "call origination" in the retail market is used to describe incoming calls. That's why you may find companies that supply virtual numbers (DIDs, DDIs, or geographical numbers) referring to these services as "call origination."

Assume you are a retail VoIP provider offering long-distance calling for your clients. You need to ensure that they can reach any destination endpoint by dialing the call recipient's number. The call recipient resides in the particular provider's network. As your clients may call to different destinations worldwide, it means that you'll need to have a connection with hundreds of networks. However, establishing a connection with each provider would be too time-consuming. That is why origination providers usually sign an agreement with one or a few operators to ensure access to the global telecom network. Companies that provide access to all destinations from A to Z are called "VoIP transit providers" or "traffic aggregators."

Transit

Transit operators work like telecom brokers by buying and selling voice traffic. The terms "traffic trading" and "minute exchange" are widely used in the wholesale VoIP market. There are many transit operators in the wholesale VoIP chain, and the more parties are involved, the lower the margin everyone receives.

Wholesale VoIP Chain

Normally, each party in the wholesale VoIP chain does not know about the clients of their clients or the suppliers of their suppliers. For this reason, from each transit provider's point of view, the company that sends traffic to them is called the "originator" and the one that accepts traffic from them is called the "terminator." In reality, only a small number of VoIP providers are purely originators or terminators. Most of them are actually trading traffic by combining origination, transit, and termination services. To give you a better understanding, here is an example of four providers in the wholesale VoIP chain:

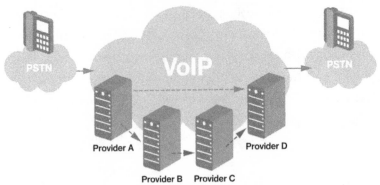

Figure 2.10. Wholesale VoIP chain.

The call flow looks like this:

- Call is initiated in the PSTN by an end-user
- Provider A collects calls from the PSTN and routes them to Provider B
- Provider B accepts the traffic and sends the call to Provider C
- Provider C re-routes traffic to Provider D
- Provider D makes the final connection with call recipient in the PSTN

As you can see, there is also a dashed line which means that under ideal conditions Provider A (originator) would send a call directly to Provider D (terminator). Theoretically, this should lower the calling rate and ensure a better quality (due to the shorter path). However, there may be many reasons why this does not happen.

Firstly, Provider D may have a minimal traffic volume requirement and Provider A may not have sufficient traffic that would reach this limit. Secondly, Provider A may not have a way to get in touch with Provider D and establish business relations. Thirdly, Provider D may have complex interconnection rules that would require time and invest-ment.

The list could go on, but let's talk about the value that the transit operator creates for its clients. Transit operators usu-ally concentrate on a few key destinations. They establish interconnection agreements with the biggest operators in those destinations and then collect traffic from different sources. By aggregating traffic volume from smaller provid-ers, transit operators can easily meet the minimal requirements that are set by big telecom companies. As a re-sult, transit operators can ensure a great offering for smaller providers: a global connection to the PSTN, a competitive price to key destinations, and a minimal prepayment that is affordable for everyone.

Types of Rate Sheets

Wholesalers usually offer three types of rate sheets to differ-entiate the price and quality of voice traffic. Each provider uses different names for their offerings, but the most com-mon are:

- 1st - Direct, 2nd - Premium, 3rd – Standard
- 1st - Platinum, 2nd - Gold, 3rd – Silver (though you may find many pricing levels like this, it's a bit outdated be-cause at this moment in time, the price of gold is higher than platinum. So it's better to use Gold-Silver-Bronze.)
- 1st – CLI, 2nd – Mixed (Blended), 3rd – NCLI (NON-CLI)
- 1st - Premium, 2nd - Standard, 3rd – Grey

The first type of traffic is typically the most stable and has the highest audio quality, but those benefits come at the highest calling rate. Providers that offer such routes usually have interconnection with the tier-1 carriers directly (this is why such routes are referred to as "direct") via the E1/T1 lines. Those who want to emphasize that calls will be terminated in the PSTN also use the names "TDM" (Time-Division Multiplexing) or "SS7" (Signaling System No. 7) on their rate sheets. TDM is a method of transmitting and receiving independent signals over a common signal path; SS7 is a set of telephony signaling protocols used to set up and tear down most of the calls in the PSTN. TDM and SS7 connections are considered more reliable than VoIP peering.

The second category is a mix of different types of routes. Mixing routes (also called "blended termination") is done by changing the proportion of "good" (high quality) and "bad" (low quality) traffic. Controlling quality influences the profitability of voice traffic; higher quality means a higher price and vice versa. All wholesalers are seeking to optimize their revenue, so the more bad traffic mixed in their routes, the higher profit margins they get. This topic will be reviewed in more detail in the "Quality of Traffic" subsection.

The third type includes NCLI, NON-CLI, or grey routes. They are the cheapest and have the lowest quality. CLI (short for "Caller Line Identification") is a feature of the direct route that allows the display of the caller's number (Caller ID) on the call recipient's phone screen. Using CLI routes means that the recipient will see the original caller's number.

NCLI or NON-CLI means that the phone screen will display the call as "anonymous" or will show a random (fake) number instead of the original Caller ID. Even though a number is displayed does not determine the audio quality, it influences whether or not the call will be picked up. Call recipients instantly pick up the call if they see names of their friends or relatives and may not answer the phone if they see an anonymous call or an unfamiliar number.

CLI routes are typically "white," meaning the termination is legal on the remote end. NCLI routes are often "grey" meaning that the call termination is done illegally, using VoIP GSM gateways. More information about grey routes will be provided in the next subsection.

Termination

Call termination refers to making the final connection of calls in the PSTN. In VoIP, there are two more common ways to terminate calls. The first one is done by signing a legal interconnection agreement with a carrier and terminating calls according to the agreed rate. The other is illegal or "grey" termination, which is organized by deploying VoIP GSM gateways in different locations and terminating calls through the local SIM cards. In this case, VoIP providers are bypassing the direct connection with a mobile operator for many different reasons that will be reviewed in this subsection.

Legal Interconnection

Interconnection is the connection between networks of two (or more) providers. It is a process that involves both technical and commercial activities. The target of interconnection is to gain access to the provider's network at the best price and quality ratio.

We've already discussed how a VoIP transit provider focuses on a few key destinations by establishing interconnection agreements with the biggest telecom operators that have the majority of the phone subscribers. Interconnection can be done with the tier-1 suppliers directly or through the lower-tier providers. National interconnections are usually done through the E1/T1 lines (such connections are also called "PSTN," "TDM," or "SS7") and enable the simultaneous transmission and receiving of multiple voice or data channels. It's important to note that VoIP transit providers are mostly using IP-based softswitches (more information in the "Softswitch Selection" chapter) to manage the call switching and routing processes, so to connect to a PSTN network, there's a need for the VoIP PSTN gateway or BRI/PRI card.

However, obtaining additional equipment, such as a gateway or telephony card, increases the interconnection expenses and complicates the infrastructure management. Because of this, some providers choose an alternative path— IP-to-IP interconnection with the international VoIP traffic supplier. IP-to-IP interconnections are more convenient for startups as they do not require any additional equipment, and the process is fast and simple.

Once the technical and commercial parts are finished, providers sign an interconnection agreement to connect their networks and exchange the telecommunications traffic. The traffic exchange can be one-way or bilateral. One-way means that one provider acts as a client (originator) and another as a supplier (terminator). Bilateral means that each of providers can be both traffic originator and terminator (such interconnections are especially common in wholesale transit).

Grey Termination

Grey traffic dates back to the nineties and exists in almost all countries that have a highly monopolistic telecommunications sector. Grey termination is an illegal workaround to direct termination with the mobile operator. Such routes have the lowest quality and the lowest price. Grey traffic is one of the reasons why international VoIP calls are so cheap, compared with the traditional termination.

The grey market usually appears because of two main reasons. First, some mobile operators are not interested in working with small players. Those operators apply a quite complex bureaucratic process of interconnection and set huge call volume requirements, too high for small providers. Another reason is the price difference between the local calling rate and the interconnection rate. (You will find a table comparing the difference between Skype and local operators' rates in the "Telecom Regulation" section of the "Understanding Telephony" chapter. It's one of the easiest ways to track the most profitable markets for grey termination.)

In countries with a liberal approach (North America, Europe), the price difference is so small that buying the equipment for grey termination does not make financial sense. It is easier and more cost-effective to interconnect with the local telecom operators. However, in monopolistic telecom markets (most of the African regions, some of the Latin American and Middle Eastern countries) the local calling rate is usually much lower than the interconnection rate. The pricing gap creates opportunities for small providers to generate revenue from grey routes. Moreover, deploying the needed infrastructure for VoIP GSM termination business is usually faster and less expensive than interconnecting with a mobile operator.

Termination with Standalone VoIP GSM Gateway and SIM Server

VoIP GSM gateway can be used for two main purposes. If a company has an intelligent PBX within its network and a few providers (a regular PSTN carrier, a GSM operator and a VoIP provider), it can configure equipment to route calls by the lowest price. In this case, PBX checks the dialed destination and, by using different routing algorithms, sends a call to the appropriate port of a gateway (VoIP calls to VoIP, PSTN to PSTN, GSM to SIM card ports).

The other purpose is to route international traffic over the Internet and terminate as local traffic on a mobile operator's network. Thus, grey route owners exploit the best on-net calling rates by terminating calls via a SIM card instead of the regular interconnection. But at the same time, call quality is usually degraded, and there is either no CLI, or else a local CLI, which does not allow identification of the original caller or the ability to return the call. For this reason, routes terminated by VoIP GSM gateway are called "NON-CLI."

There are two types of VoIP GSM gateways: the standalone GSM gateway (with ports to insert SIM cards) and the SIM server-based gateway (control of SIM cards is done by SIM server). Deployment with a standalone VoIP GSM gateway (see picture below) is used for smaller deployments and is inexpensive compared to the second option.

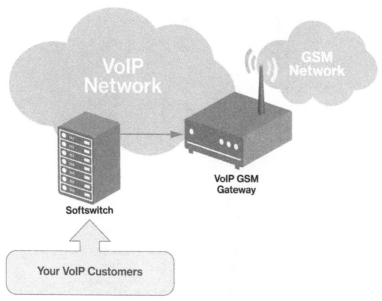

Figure 2.11. Standalone VoIP GSM gateway.

The SIM server (also called "SIM bank," "SIM box," or "SIM array") is a device which holds SIM cards that can be accessed remotely. The main idea of a SIM array is to connect separate gateways with GSM antennas or modules and SIM cards. All SIM cards are in one place while VoIP GSM gateways can be distributed in different areas. Such deployment allows the insertion or change of a SIM card remotely on the VoIP GSM gateway any time and enables the centralized management of a call termination business (see picture below).

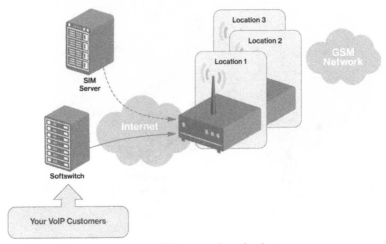

Figure 2.12. SIM server-based solution.

The main benefits of the SIM server-based solution compared with standalone VoIP GSM gateways are:

- **Efficiency in changing SIM cards.** If standalone gateways are located in different places, you cannot replace SIM cards fast and frequently. However, with the SIM server, it is done from one physical location and it is easy to change a SIM card whenever it is blocked.
- **Centralized management.** If you are using more than one gateway, it is hard to change their configuration in the same way at the same time. However, when using the SIM server, gateway management is centralized.
- **Reduced SIM blocking.** If you are using a standalone gateway, you cannot use the same SIM card at different locations spontaneously; it needs to be moved physically between gateways. The standalone gateway uses few SIM cards in one location to terminate lots of calls simultaneously. Such traffic is significantly different from

the normal "human usage" of SIM cards and can be easily identified by a mobile operator. Such call termination activity is treated as abuse and SIM cards are blocked very fast. On the other hand, using the SIM server allows you to automatically change the "location" of the SIM card by simulating "human behavior." As a result, it reduces the SIM blocking rate because it makes it harder for mobile operators to distinguish between a call termination machine and real human behavior.

Some of the suppliers that deliver VoIP GSM gateways and SIM banks are: Hypermedia, Antrax, Sysmaster, Topex, HyberTone, iQsim, Eurotech Communication, 2n, Portech, and Simberry (created by co-founders of iQsim, another VoIP GSM and SIM bank solution provider).

Fight Against Grey Traffic

Deploying a VoIP GSM gateway or a SIM server for commercial purposes (VoIP termination or origination) is illegal because there is no direct contract of interconnection with mobile operators. Despite this, grey route providers still choose this business model because it is quite profitable: the investment is relatively small, and it allows offering international traffic at the same rates as the local calls.

Of course, there are also companies that sign legal interconnection agreements, but they are usually much larger than those wholesale traders who deploy SIM servers. Mobile operators are fighting against such "fraudsters" using various methods of blocking SIMs. Despite this, revenue from grey termination is growing each year. (It's important to note that there is a niche for grey termination only in countries that have a telecom monopoly.) Recent surveys show that on average 15% of mobile providers' revenue is lost to VoIP providers using SIM boxes.

There are laws to fight against the termination bypass and sometimes you can even go to prison for organizing such activity. In addition to this, mobile operators use anti-fraud systems to detect grey traffic and turn off the SIM cards. In general, detecting the fraud is based on analyzing various calling parameters. These are the grey termination indicators:

- **Incoming vs. outgoing calls.** There are more outgoing calls than incoming.
- **Called destination.** Almost all calls are on-net (terminated within mobile operator's network).
- **The timing of calls.** There is a high traffic volume from the same SIM card. Moreover, the time gap between the end of one call and the start of another call is very small (near calls).
- **Location.** High traffic volume from the same location (it is common for standalone VoIP GSM gateways as they are deployed in a fixed location).
- **Called numbers.** Called destinations change quite often, so the number of called destinations divided by the number of calls is close to one. For regular subscribers, such a ratio would be lower than one because normally each person has a few people (family and friends) whom they call the most often.

Quality of Traffic

Voice traffic is often measured by quality: the higher the quality, the more it costs. This is why managing voice quality is so important to ensure optimal profitability in a wholesale VoIP business. This subsection reviews the main quality benchmarks and demonstrates how to see if your supplier is intentionally trying to lower the quality.

Filling the Pipe with "Good" Traffic

The service provider's softswitch can be treated like a pipe with certain limits. (These limits will be discussed in more detail in the "Softswitch Selection" chapter.) If this pipe is filled with good traffic (high quality calls), the provider will get the highest profit. If the pipe is flooded with bad traffic, it will result in lower profit. Moreover, bad traffic increases the load on the system, which affects its performance and can cause quality-related issues, or even system crashes.

Good quality calls are usually generated by consumers and business users, which is why transit operators are constantly looking to cooperate with retail providers or try to approach bigger corporate clients directly. Some traffic aggregators, at a certain stage of their business, open a unit for retail VoIP business. This allows them to create an additional source of income and have complete control over the call origination quality.

What happens with all the bad quality calls? There is a niche market—such calls are called "dialer traffic" and are usually generated by call centers and telemarketing agencies. Those companies have many operators, who are continuously making calls and a majority of the conversations are short. It causes a heavy load on the server, so providers that handle this type of traffic must have a robust infrastructure and transmit huge call volumes to obtain reasonable profit margins.

Quality Benchmarks: ASR, ACD, and PDD

To determine the route quality, providers use specific benchmarks, such as ASR (answer seizure ratio), ACD (average call duration), and PDD (post dial delay).

ASR shows the percentage of answered telephone calls with respect to the total number of calls attempted (seizures). ASR is vastly dependent on end-user behavior, but in general, a higher ASR percentage represents a better call quality.

ACD reflects the average length of calls and is usually measured in minutes. If customers stay on the phone longer, it is assumed they are satisfied with the audio quality. On the other hand, if the quality is terrible, the call recipient hangs up very quickly, which results in the low ACD.

PDD is the time between the start of the call (dialing the last digit of the called number) and the moment the phone of the called party starts ringing. A low PDD shows that a connection occurs very fast, which results in a positive user experience. Consumers are generally used to a short PDD in the PSTN and usually react negatively when there is a long pause after dialing.

Some providers that have low quality termination with high PDD use a fake ring feature. A fake ring is generated by the softswitch; as a result, the caller hears a ringing signal (beep-beep) instead of silence until the call recipient picks up the phone. It may create confidence for the caller, meaning that they stay on the line (which is the aim of the provider). But in general, it is considered bad practice because the fake ring hides cases where the call recipient is unreachable or busy. In such situations, a caller will still hear the ringing signal and may be confused.

False Answer Supervision (FAS)

FAS is one of the causes of low quality benchmarks. In general, FAS refers to the situation when the billing of a call starts earlier than when the call recipient actually picks up the phone. This results in the incorrect overbilling of calls. Fake FAS service is sometimes used in wholesale VoIP to add false billable airtime to the caller by simulating the messages of a real mobile operator. The calling party is being charged while hearing messages like "the call recipient is out of coverage," "we are sorry, but the party you are calling is unavailable at the moment," or "the number you are calling is not reachable at the moment; please call back later." Playing such messages does not cost anything for the provider. The message is uploaded to the softswitch and the routing is configured in a way that a part of the calls is sent to the IVR (interactive voice response) system, which is set to play the recorded file.

I remember my first call with FAS very well. I was calling a potential client for the first time. Someone picked up the phone and said "hello?" I introduced myself and, after a short pause, I heard "hold on a minute." After that there were many different background noises like someone going upstairs, a dog barking, a woman asking for something (I thought it was his wife). To me, it sounded like a natural situation, and I stayed on the line for a few minutes, waiting for this "potential client" to come back. My patience lasted for around four or five minutes before I hung up the phone.

Back then I did not understand that this was FAS, but after I had heard the same message three or four times, it became clear that I was repeatedly getting a recorded message. When I studied more about FAS, I was slightly shocked that someone could apply such methods to generate revenue. But the more I spoke to other wholesalers, the better I understood that it is a quite common case in the VoIP market. Even though FAS is considered fraudulent and may be illegal in many countries, there are still many providers that mix fake FAS calls into the real traffic to make extra money. Mixed traffic is a "public" secret; it's just that no one reveals the proportion levels.

Fraud in Wholesale

Fraud is a common activity in the wholesale VoIP market. It is a process where someone makes a profit at another's expense, illegally. Compared with VoIP hacking, which requires good technical skills, VoIP fraud is about being able to persuade and convince other people through the use of various manipulation techniques and tricks. The worldwide cost of VoIP fraud over the last few years has reached billions of dollars and is still growing. This subsection reviews different fraud schemes, so you can easily identify and prevent them.

Fraudsters first create a fake company and personnel with fake names. Then they try to find some inexperienced companies (usually that are new in the VoIP business) and sign false contracts with them, so it would seem that everything is done legally. After this initial process, various fraud schemes can follow, leading to huge losses to the fraud victim. Sometimes fraudulent actions affect a bigger chain of providers. When a company unexpectedly loses money, it cannot pay bills to a supplier, and then the supplier cannot pay bills to their supplier, and so on. Some of the tricks used in fraud schemes are listed below (you may find more examples at Voipfraud.net).

Excessive Trust

First, a fraudster must earn the trust of his victim. This is done by simply making payments on time, after signing a contract. As in any other business, this is accepted as a good sign, demonstrating a positive credit rating to a client. However, the real aim is to switch from a prepaid to postpaid payment mode on the same credit limit. Once that credit limit is reached, payments are still made on time until the fraudster has earned a really high credit limit that allows more money to be made from fraud.

Fake Payment Confirmations

When starting new business relations, it is natural to trust each other, and this trust can increase or decrease after some time. Once fraudsters are credible enough to get a big credit limit, they move to another trick—win more time by sending fake payment confirmations. If payments do not arrive on time, it is easy to blame a third party, the bank. Sometimes sorting this out can take days or even weeks. By managing this process, fraudsters can evaluate the level of trust measured in time, e.g. how many days they can win.

Fake Disputes

Disputes in wholesale due to mismatched call detail records

(CDRs) are a normal process. However, fraudsters use this to negotiate the possibility of prolonging their period of using credit to make calls. As an example, they can send angry emails claiming that part of the CDRs do not match, or there are a lot of low quality calls while, according to the agreement, the quality level should have been higher. Of course, in reality there is no real reason behind this except to win more time. Usually, more vulnerable providers continue accepting traffic from them during the disputes, which can lead to big losses.

Friday Night Request

VoIP providers have the majority of their personnel in the office during regular business hours on working days and some employees (usually engineers or network administrators) who are responsible for system monitoring 24/7 (24 hours per day; 7 days per week). By knowing such schedules, fraudsters pick a moment (when there's no manager) to present themselves as a well-known client to the network administrator. As a request, they ask to add a new IP urgently and pretend that in some other way they will be punished if the network administrator will not help them. If there's no strict policy about such situations in a victim's company, and such a scheme works, then the network administrator adds the requested IP to the softswitch configurations. It becomes worse if the IP is added with a big credit limit or no limit at all.

On Monday morning, when managers come to the office, losses can be so huge that such a provider can be out of the business the same day.

To fight against such dishonest people, there are various informational portals (Voipfraud.net), forums, and LinkedIn

Groups where people can share information about fraudsters (the company, the name of the person, and an explanation of the situation). Also, to get more trust and confidence in companies, it is recommended to ask for references.

Chapter 3.
Deciding on a Business Model

As you are now familiar with all possible VoIP services, the next step is to decide which business model is right for you. To answer this question you need to understand your strengths, choose the risk and responsibility level, make an analysis of the competition and your target audience, evaluate telephony trends in your market, and summarize everything by financial calculations.

Your Strengths

Running the business requires having an entrepreneurial spirit, special skill set, knowledge, and experience. Many people consider providing VoIP services, but the most successful businesses are established by those who are aware of their strengths and weaknesses. They pick an area they enjoy, where they can leverage their skills.

Successful VoIP entrepreneurs have had some previous experience in the telecom industry; many are former employees in VoIP companies. By working in VoIP companies, they were able to understand the key business operations, establish contacts with potential clients, and gain all the other practical information that has helped them to build confidence in their own business. You can get the basic know-how by reading this book, but the best experience will come once you start practically applying that knowledge in the business environment.

Many entrepreneurs establish a sole proprietor company when starting a VoIP business and take care of all operations by themselves. This provides complete freedom and allows fast decision-making, but one person cannot be great in all areas. Businesses run more smoothly if operated by people who are experts in their own areas. A wide variety of VoIP skills are required to manage a startup company, but with a small VoIP business, there are two major skill roles—technical and commercial.

Technicians are responsible for choosing the reliable infrastructure, managing it in daily work, and solving technical problems. They have system administration and troubleshooting skills, as well as a good understanding of IT.

Commercial people deal with sales and marketing processes, organize negotiations with clients and suppliers, and manage relations with business partners.

It would not be difficult for you to pick an area that plays to your strengths. For example, even though I love technology, I'm still more a commercial person because I like communicating with people; I could not spend the whole day in front of a computer troubleshooting issues or running general operations.

Here are questions and comments that will help you understand your strengths and weaknesses better. Answer them for yourself:

- **Experience.** How much experience do you have in VoIP? Without much experience, what will be your competitive advantage against other providers? How will you compensate the lack of experience if clients expect a high level of competence?

- **Target market.** Do you have an existing target market? Those starting a VoIP business should have some existing clients or at least a few "beta" customers who are waiting for the launch of your VoIP service and are willing to share their feedback with you. Direct access to this target audience is a big advantage because people already know and trust you. As a result, selling a new VoIP service will be much easier. If you do not have access to the target market, how will you create a new demand in the market? Or how will you encourage other clients to switch suppliers and use your service?

- **Services.** What type of service do you want to offer? If you have experience with corporate clients, think about services like SIP trunking or virtual PBX. If you work with an ethnic community, consider calling cards, pinless dialing, call shops, or a similar service that will help them to call their home countries at a low cost.

- **Resources.** Which of these resources do you have: money, technical skills, or sales skills? If you have money but no experience, consider being an investor in a VoIP venture. There are many experienced entrepreneurs willing to start this kind of business, but lacking enough cash flow to launch the service. If you are a technical person and know how to install, to acquire, and maintain a VoIP infrastructure, find a partner who knows how to attract potential customers and convert them into long-term clients. If you are a sales person and have the right contact and potential users, think about hiring a partner who will take care of the technical work of organizing a reliable infrastructure to launch a VoIP service.

If you have already answered those questions for yourself, the next step is to choose the level of operation. You will see that if you are good at selling, you can become a reseller and find a provider that will arrange the technical environment, back-end support, and will pay reasonable commissions for each client you bring. The next section explains the difference between reseller and service provider.

Reseller vs. Provider

There are a few ways to be involved in the VoIP business. The first one is starting the service from scratch—buy your own system, choose suppliers and become a provider. This gives you full control and flexibility, but it requires a sufficient cash flow and is especially challenging if you do not have any previous experience in VoIP.

The second option is choosing a reliable provider with an established infrastructure and becoming their white-label reseller. In this case, you do not need any initial investment and your core responsibility is to market and sell a VoIP service under your own brand. This is probably the best way in for those who do not have any technical skills but would still like to work in the VoIP market.

The last option is to become an agent and work on commissions as a non-contractual employee. Agents usually work under a well-known and trusted brand in their market. This is probably the easiest method, but because it is not so different from being an employee of the company, it will not be reviewed in detail here.

Resellers

Resellers are usually freelancers or companies that have some other core business; VoIP is not their main activity. They work under their own brand (white-label) and buy voice traffic and DIDs from the main provider at wholesale prices. Resellers have their own customers and apply individual tariff plans that are different from the core provider. However, there are also cases where providers and resellers use the same tariff, and resellers receive a fixed profit margin from their sales. This is especially common among distributors who work with physical VoIP products, such as calling cards or printed top-up vouchers. By distributing calling cards with a printed value, they get either a fixed amount or a percentage once a card is scratched and the PIN entered.

Resellers normally work in their local market or territory. If they perform well, the provider may give them exclusive selling rights for the specific region. Resellers deal with end-users or subresellers (if such a capability is allowed by the provider) and take care of operations, such as pricing, charging and collecting money from clients, handling invoicing, monitoring calls, setting up new accounts, and presenting their services as if they were the owner. Having such sales partners is quite popular in all VoIP business models since it saves time in dealing with end-users and allows a scalable income by sharing the revenue. Paying commissions is usually more preferable than hiring, training, and paying a fixed salary for a full-time employee. Even in wholesale there are many cases where traffic aggregators deal with resellers because they can get quality retail traffic and ensure a bigger profit margin.

One of the main differences between reseller and provider is the ability to manage routing to multiple voice traffic suppliers. Providers use their own softswitch with a few call termination suppliers and have the control of the call routing, whereas resellers do not have their own infrastructure and manage their operations through the reseller portal supplied by their provider.

Sometimes resellers have two or more suppliers, but managing a business through multiple portals is a time-consuming task, and there is a better solution. Some providers offer an additional service: a switch partition that allows resellers to add their own carriers by paying a small monthly or per-minute fee. But this is a rare practice because VoIP traffic is the main revenue stream for all providers and they are not interested in creating competition for themselves.

Below you can see a comparison table which reviews the advantages and disadvantages of being a reseller or provider:

	Reseller	Provider
Investment	Resellers work on a commission basis by using a provider's infrastructure. As a result, there is no need to invest anything or to acquire equipment. If resellers operate on a prepaid level, they need to buy some traffic up-front. On a postpaid basis, they have some credit lim-	Providers require a bigger initial investment to acquire needed infrastructure and equipment to operate the business. A larger company incurs other regular expenses, like office rent, employee salaries, and up-front or postpaid pay-

	its and pay weekly or monthly to their provider. This allows an easy start to the business with the minimal investment and risk.	ments for the voice traffic.
Technical knowledge	Resellers don't need any VoIP technical knowledge and can focus on marketing and selling activities. Technical operations are provided by the provider.	Providers maintain a complex telecom infrastructure that is operated by experienced engineers with good troubleshooting skills.
Control and flexibility	Resellers cannot fix call quality-related problems because they use a provider's equipment. Moreover, prices can be changed by the provider, which puts the reseller in a difficult situation because there may be no alternate supplier that could be chosen instead. Lack of control over pricing, quality of calls, and service portfolio may have a negative impact on	Providers have their own infrastructure and a full control of the business. Moreover, providers have interconnections with few traffic suppliers, which allows routing calls through the most profitable path. If one of the suppliers increases calling rates, it is easy to route calls through another traffic provider. Al-

	reseller's business. The possibility of more freedom, flexibility, and control of the business is one of the key reasons that re-sellers want to become service providers.	so, providers can expand their service portfolio if they see a demand for a new VoIP service or so-lution.

Reseller Program

To attract and motivate resellers, providers must have a solid program to get interest from potential sales partners and make existing ones loyal. The most important parts of a partner program are: commissions, marketing, and training (both commercial and technical).

Commissions determine how much the reseller will earn from each sale. Commissions can be provided for either the successful referral or the closed sale. The first case refers to an online affiliate program where a partner directs visitors to the provider's website and receives commissions based on the orders that are processed by the referrals. But the majori-ty of resellers get paid for the sales they close. The profit margin may also depend on sales volume, so resellers are motivated to sell more. Sometimes it makes sense to use sales targets and quotas, but this is usually applicable only to the highest level of resellers who have exclusivity in their market.

Marketing allows a reseller to get qualified leads. There are different kinds of marketing methods: marketing can be done by the reseller, by the provider (especially if the reseller acts under the provider's name) or by shared efforts; in this case, the reseller and provider cooperate to create the mar-keting strategy.

The partner program includes materials about products and services, pricing schemes, webinars, promotions, monthly newsletters, and other helpful information. Sometimes there is online or personal training available to develop the necessary commercial and technical skills for a reseller. Providers usually offer some web-based tools that allow the running of live reports, the managing of customer and subreseller accounts, the reviewing of sales and marketing material, and access to a help desk for support and assistance.

Choose the Level

There are three general levels in a VoIP service chain: customer, reseller, and provider. Here you can see a simplified call flow which shows the parties involved in an international call:

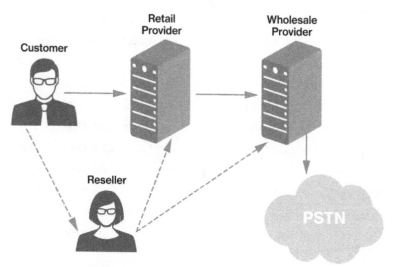

Figure 3.01. Call flow parties: customer, reseller, retailer and wholesaler.

The customer (end-user) initiates a call. It comes to the retailer's equipment; from here it is passed to a wholesaler, and finally the call is terminated in the PSTN. As you see, there is a reseller who may also be involved in the chain. The lines pointing to it and from it are dashed because a reseller does not have any physical equipment. Both the retailer and wholesaler act as providers because they have their own infrastructure.

When deciding on a business model, a simple rule is to begin with the smallest step. Get a feeling and understanding of the challenges of this level and then move to the next one. Such a process takes longer than jumping to the top from the beginning, but it ensures minimal risk, increases the success rate, and provides a smooth transition between levels.

If you do not have any experience with VoIP, become a VoIP user yourself and analyze the services of multiple providers from the client's perspective. How satisfied are you with the service provider? Do they meet all your expectations? If you start to notice points which you could improve upon, try to communicate this to your provider. What is their reaction? Are they client-oriented? Do they fulfill your requests? The idea is to choose one provider that you are very happy with and would definitely recommend to others. If you feel like an ambassador of their services, ask if you can become their agent or reseller. A majority of providers would agree because this provides another sales channel for them. Then you can move to the next step—becoming an agent or reseller.

Becoming an agent or reseller is a great chance to start earning from VoIP services by taking minimal risk. No investment is needed at this stage. You can concentrate only on one specific process—selling. All other operations are done by the service provider.

However, after that initial stage you also start taking on more responsibility. Once you've acquired some clients who use your services, they'll have different expectations from you. If they are not happy with something, they will blame you, and you'll need to address those issues to the service provider. In the end, you'll spend time not only on sales but also on customer services to retain your clients. Keeping a certain level of client satisfaction is the key element to revenue growth. After you gain some experience as a reseller, ask yourself if you could do something better than your current provider. Are your ideas compelling enough for your clients to switch to another supplier? Once you find a compelling reason to make a change (provide lower rates, increase call quality, make the overall process simpler, or offer a better client service), you can communicate this to other users and see how they react. Are they affected in the same way as you are? Or this is just a "nice to have" for them? Under ideal conditions, you should find "must-haves"—reasons that would encourage clients to migrate to another supplier. Once you've found a competitive advantage and there is a group of people ready to follow you, it is time to move to the next level.

Becoming a service provider is the highest level. It requires the biggest investment; you'll take on a lot of responsibilities, and most probably you will need to learn a lot of new things and gain new skills. While it may sound like a big challenge, it opens a completely different level of opportunities. You will be an independent supplier who can take complete control of your business. You will have the possibility to earn more than you did as a reseller. As a bonus, you will feel like making bigger changes, not only in your personal life but also in your work where you will be creating additional value for your clients.

Market Research

Now, as you are aware of your strengths and understand the differences between the levels of opportunity, it's time to talk about market research. This is not some trivial process or an idle curiosity to indulge in before establishing a business; yet a lot of startup companies do not even bother with it. However, it is very useful to see the wider picture and make more rational and focused decisions about your business. In this section, you will learn more about your target audience, your competition, and ways that your curiosity can help with market research.

Target Audience

The first thing to do is identify your ideal client. As a VoIP provider, understand who your clients are, what their biggest needs are, and how you can fulfill them. Why should clients choose you? A very common case among many entrepreneurs is that they first acquire the needed infrastructure for a VoIP business and then hope to deliver services to everyone that needs VoIP. Unfortunately, that's a big mistake which can lead to a speedy failure.

The best way to start is to choose your target audience based on your experience with the people you are used to working with. Those can be your clients in your current or previous job or members of a community, group, or organization you belong to. As an example, businesses of calling cards and call shops usually belong to people from the ethnic community. Such business owners are aware of the ethnic culture and know how to communicate with their clients in the right way.

Many entrepreneurs tend to start their own business after learning and getting necessary experience in their previous company. This seems natural in each industry, and VoIP is not an exception. The growth of a person in the business environment is similar to the transition to another level (e.g. from agent to reseller, or from reseller to provider). By working in the VoIP provider's organization, employees become aware of the business processes, understand how to deal with clients and suppliers, see the challenges of the company from the inside, and get insights on how to improve this. As a result those who feel they can make something better, start their own business.

The perfect situation is to have an existing customer base to whom you can easily introduce a new service. Probably the best examples of this category are ISPs and wireless Internet service providers (WISPs). They can easily add VoIP to their service portfolio. Depending on the competition level, VoIP services can be a source of additional income or at least a way to increase the client retention rate.

Recently, I have had an interesting case with an ISP in Alicante, Spain. They were terminating almost half a million minutes per month, but all of those calls were without any profit. Their strategy was to give a package of VoIP minutes, a virtual number, and equipment (an ATA adapter and router) for free if the client signed a two-year agreement. As a result, clients received a significant value free, while the service provider generated profit by selling Internet services.

A very similar case was with a WISP in Naples, Italy. They served only business clients and were approaching them with a packaged offer that consisted of an Internet connection, local and international calls using VoIP, virtual numbers, and cloud PBX. Even though the Internet was their core service, adding extra value gave them a lot of space to negotiate with clients and show a competitive advantage. This WISP could easily offer calling services with a negative margin and generate a profit from Internet connectivity services or cloud PBX.

To summarize both of those cases, having multiple services in one package provides great flexibility.

If you are starting a VoIP business from scratch, the process is more complicated, but solutions still exist. In the beginning, determine whether there is a niche market which you know much better than others. The majority of markets are already overcrowded with VoIP offerings. Competing there without experience would be a complete disaster. But if you have established relations with a specific target audience or social group, or have maybe worked in a niche industry or association, you could use your experience to develop a different approach from your competitors. Moreover, knowing the market better than others and having the right contacts also reduces the customer acquisition cost.

One such success happened to a client who was serving call centers in Tirana, Albania. He was a former call center manager, so he knew the main challenge of his target audience very well. There was no affordable call management tool that was simple for agents to use and worked well at a high frequency of short duration calls. When he found the right solution, he proposed a bundled package for the call centers: a call center management system with SIP trunking services. His customers were happy with the bundled offer and treated their provider as an expert; word-of-mouth spread and resulted in a rapid growth of that VoIP business.

Another good example is a company that provided payment gateway services for the Middle East market. The gateway processor had a facility to review all financial transactions of their clients, and part of them were retail VoIP providers. After noticing numerous payments for telephony services, they decided to establish a new VoIP company. Even though they did not serve the retail market directly, they had needed information, which helped them to develop a good marketing and sales plan.

Competition

Friends often ask, "How can someone use services of some small VoIP providers when there are applications like Skype and Viber?" I would probably ask the same question if I wasn't aware that those small providers can attract clients by offering even lower international calling rates and retaining the same call quality as well-known applications. However, competition between small and big players is a very strong factor that requires your special attention.

After you know your target audience, it's time to research the top service providers that are used by your potential clients. Also, you can review those VoIP suppliers that, while they exist in your market, your target clients do not use their services or are not aware of them at all. It helps to view the picture from the customers' perspective. With whom are you competing? Are they leading local telecom operators, worldwide OTT applications, or small niche providers? Is there enough market space for another provider? Will you try to encourage clients to switch from their current provider or will you try to approach those that are not aware of VoIP services at all? All those questions must be answered before taking any further step.

Do not be afraid to compete with the big players. For a small business to survive and grow, it can be enough to cover as little as 0.1% of the local VoIP market. The leading providers think globally and cannot meet the expectations of all their clients. As a result, part of their customer base is not fully happy with such an approach, and looks for alternatives. This opens great opportunities to the niche VoIP service providers that can demonstrate their value and make a positive impact on their clients' businesses.

Curiosity

Most entrepreneurs are curious people, and this quality helps them dig deeper into the area that they enjoy. If you are curious, here are some recommendations for you to do as market research, which should be an integral part of your business.

First, learn all you can about the marketplace where you plan to operate. Talk to potential customers and try to understand what drives them to use a specific brand, how loyal a client they are, and what the main reasons are for them to change suppliers. Then analyze the competition by pretending to be a potential client of their services. Try to look at the service from the client perspective. It helps to understand what can be improved.

Next, read VoIP industry magazines, blogs, articles, and books. If possible, attend the telecom exhibitions and tradeshows, even if they are not organized in your country. This gives you a global view and helps to understand world-wide trends. If you cannot afford to travel, there are many forums and social groups that connect VoIP professionals, especially on LinkedIn. Join those groups and ask the right questions; experienced people will always be willing to help you. There are also special websites, like Quora, where you can post questions and have industry experts share their feedback on this topic.

Finally, test the waters before making large commitments of time or money. You'll learn how to do that in the "Launching Your Service" chapter.

Chapter 4.
Softswitch Selection

You've already read the term "softswitch" in this book many times. It is the core element in a VoIP provider's environment, so it is worth a review in this dedicated chapter. The ancestor of today's softswitch was a switchboard (reviewed in the "Evolution of Telephony" section of the "Understanding Telephony" chapter). First it required human operators to connect calls. Later it was automated, and finally, after the invention of the transistor, the switch became a computer-based electronic system. There are still many hardware-based switches, but softswitches are becoming more popular due to price and flexibility. They are widely used in the VoIP market as they turn any computer or server into a powerful telecom switching machine. This chapter explains the softswitch components, main functions, and the main principles for choosing the right softswitch.

Softswitch Components

Current softswitches are divided into two categories: Class 4 and Class 5. Class 4 are dedicated for wholesale transit; their main function is the routing of large voice traffic volumes. The most important characteristics of a Class 4 softswitch are: system performance, integrated billing, flexible routing, codec and protocol conversion, and comprehensive reporting. Class 5 softswitches are used to provide local and long-distance calling services for end-users. The key functions of a

Class 5 softswitch are: PBX features (call transfer, call parking, call forwarding, follow-me, hunt groups, IVR, and call queues), different types of authorization, flexible rating, and support of prepaid and postpaid accounts. Moreover, both classes of softswitches may have many extra features that can be inbuilt or come as extra modules (also called "add-ons," or "plug-ins").

There are also many cases, where VoIP providers offer both retail and wholesale services. The best option, in this case, is to use a single solution that incorporates both Class 4 and Class 5 functions. Such softswitches are especially popular among small companies that want to minimize the investment and have an easier administration of a single system instead of managing two independent platforms. However, if the business grows and voice traffic reaches a certain level, it is recommended to split the system and deploy separate softswitches for each segment. Handling big traffic volumes requires high system performance, and if the server is overloaded, it can result in call quality issues, improperly working functions, or even system crashes. It is not a big problem for wholesalers since they can automatically route calls to another termination partner. However, end-users are especially sensitive because they usually rely on the phone service of a single provider. To avoid this, VoIP providers should constantly monitor their systems and make decisions about expansion before problems start affecting their clients.

Softswitches consist of some key components that can be considered as the main frame of each VoIP business. In Class 4/5 softswitches, all components are available in one system, but in bigger deployments they may run on separate servers to ensure better performance.

SIP Server

It is the core component of a softswitch or IP PBX. A SIP server is also referred to as a "SIP proxy." A SIP server mainly deals with the setup of all SIP calls and provides basic call control features, such as call forwarding, call waiting, and call transfer. However, a SIP proxy does not transmit or receive any audio.

RTP Server

Also known as a "media server" or "RTP proxy." It is responsible for the multimedia, such as audio, video and data transferring.

Billing Server

It communicates with VoIP gateways, SIP servers, and other elements of a VoIP network by providing authentication, authorization and accounting (AAA) services. It also gathers billing information, rates it accordingly, modifies customer balances and creates transaction records. (More on billing functions in the "Billing" subsection of the next section.)

GUI Server

GUI refers to a "graphical user interface" and is also called a "web server." It provides browser or server-mode management for configuration and maintenance of the system for various access levels (administrator, accountant, reseller, and user). Without GUI, system management would be very complicated and possible only with a good technical background.

Database Server

The database (DB) server is the storage center for operation data, configuration parameters and user information. There are various types of databases, such as MySQL, Oracle, Microsoft SQL, or PostgreSQL.

Application Server

The application server provides various IP PBX or Class 5 features, such as voicemail service to subscribers, audio conferencing, interactive announcement messages and unified messaging.

Load Balancer

Used in a cluster environment to distribute calls among multiple SIP proxy servers depending on the load each server can handle. Even if one of the SIP servers is down, the load balancer can notice this and stop sending calls there. A load balancer helps to achieve the optimal system performance, maintain high availability of the services, and enable a scalable architecture.

Voice Gateway

Provides the connection between different networks (usually VoIP and the PSTN), translates the multimedia flow, and transfers the signal. Usually, it is a third party device, such as a VoIP PSTN gateway or a BRI/PRI card, but there are still some hardware-based switches that have an integrated voice gateway function and allow access to both VoIP and the PSTN networks.

Softswitch Functions

Softswitch is a robust program with a multi-feature set. The key functions are billing, routing, reporting, and monitoring, but it does not mean that each softswitch is limited only to them. Functions can be an integral part of a softswitch, or partitioned off into modules. Such system structure is called modular architecture and enables different add-ons, depending on the client's requirements. This section explains the main and additional softswitch functions.

Billing

VoIP billing can be a part of a softswitch as well as a completely separate platform. Small providers tend to choose all-in-one softswitches with an integrated billing function as they are more cost-effective. Others prefer to use a distributed environment—a standalone billing system that retrieves data by interacting with a softswitch or a SIP proxy server. Such implementation helps in achieving a higher performance. Because of this, it is often used by wholesale providers that have bigger voice traffic volumes.

To understand why billing is considered a key function, here is an example of billing-related tasks that are done after the call is completed. Whenever the user initiates a call, it is passed through the provider's switching platform which generates a CDR. It contains the attributes of time, duration, source and destination numbers, and completion status. CDRs are later used in all processes that require information about calls. One of those processes is invoicing which is usually done once per month in a postpaid business. To prepare an invoice, it is important to have the information about charges for the total amount of calls during a month. The calling rate depends on:

- Call status (only answered calls are charged)
- Call duration (the longer the user talks, the higher the cost)
- Rounding (calls can be billed per-second or per-minute)
- Destination number (usually the rate is defined by the prefix)
- Subscription plan for the specific user (whether it is a flat rate or a usage-based tariff)

Also, there may be a different rate depending on the hour of a day (on-peak or off-peak) and the day of a week (workday, weekend, or holiday).

Between all of these we have a bunch of changing parameters. All of them should calculated to complete a billing process and prepare an accurate invoice. If the provider has a very simple billing method and a small user base, it is still possible to perform this manually by playing with the formulas in Excel, but in most cases, it is a big headache. Manual work is time-consuming and long hours can cause mistakes that lead to losses due to inaccurate calculations, disputes, or other negative consequences.

Therefore, it is much better to avoid manual work by automating the billing process to save a lot of time, energy, and money. A robust billing platform can process large volumes of voice traffic without affecting other business functions, thus generating invoices accurately. Such a facility allows focusing on the core business activities and automating the recurring processes.

Billing consists of additional elements, such as authentication, authorization, accounting, and rate management.

AAA

Authentication, authorization, and accounting is a method used for caller identification, giving appropriate access to the softswitch, managing permissions, controlling the service usage, and providing billing information. AAA functions are usually inbuilt or provided by a RADIUS (remote authentication dial-in user service) server. Tasks related to the CDR retrieval can also be performed by FTP (file transfer protocol) or SQL (structured query language) servers.

Authentication provides a way of identifying a caller by checking their credentials and comparing them with the information stored in a database. If a user is not authenticated, the system automatically denies access. There are two main methods to authenticate users: by IP address and by username/password. IP authentication is widely used in wholesale for IP-to-IP interconnections and, in some cases, in the corporate market to authenticate PBXs. Username/password authentication is more common in retail VoIP and is used to identify SIP clients (IP phones, mobile dialers or softphones).

The subsequent process is authorization, which determines whether the authenticated user has the permission to perform specific activities or to get certain resources. In most softswitches, it is set by default that all users are authorized to call internally for free (this includes "pure" SIP calls, such as extension-to-extension or computer-to-computer). External calls to the PSTN can be allowed or blocked to prevent VoIP hacking (especially calls to premium numbers, special service numbers, or expensive destinations, such as the Gambia, the Maldives or Cuba).

Other examples of authorization are automatic number identification (ANI) and authorization by PIN (reviewed in the "End-user Services" section of the "Choosing a VoIP Service" chapter). Both methods are used in services that require calling through an access number. ANI, or pinless service, is more popular; it allows dialing in a more convenient and faster way. PIN authorization is not so common but is more secure.

In addition to this, authorization allows control of the balance of a prepaid user and the credit limit of a postpaid client. Not all systems are capable of working in a prepaid mode, so it is very important that softswitch supports both types of accounts. A prepaid account ensures that the user will not spend more than the balance; postpaid means that the user will not exceed the credit limit. Having both prepaid and postpaid in the system allows easy control of the payment process and avoids possible losses.

Accounting is the final part of the AAA framework, which measures the resources a user has consumed during a specific period. Accounting allows you to keep track of payments, generate invoices, enable multiple currencies, and manage multi-level taxes. Payments can be handled in a few ways: cash, voucher purchase, or online transaction by using a credit card or e-banking. Online transactions are usually processed by payment gateways, such as PayPal, Authorize.Net, MoneyBookers, Ukash, or 2CheckOut. An integrated payment processor allows quick and convenient payment from the user panel or the provider's website and automates the payment and service delivery processes. Each processed payment is reflected in invoices that are usually generated and sent to users once per month. In wholesale it is also popular to apply shorter invoicing periods due to the higher cash flows. One of the common invoicing methods in VoIP transit is called "7/3." It means that all traffic usage of one week (7 days) will be invoiced on Monday of the following week and should be paid within three days.

Most accounting tasks are done in the softswitch, but sometimes providers use third party applications if the softswitch does not support a required invoice template or if the local laws require a complex tax management.

Rate Management

Rate management involves work with the periodic and one-time packages, tariffs from different carriers, subscription plans, and other fees. Rate management in wholesale is quite different from retail—both cases will be reviewed in more detail.

Rate management in the VoIP transit business involves activities such as importing tariffs from new vendors, creating offers for potential clients, updating prices of current providers, and sending rate updates to customers. Usually, every single vendor has their own tariff layout and format. Such tariffs must be converted to a "normal" format that is understandable to the system. Once a rate sheet is converted, the file can be imported to the softswitch.

The next step is to define the relative or absolute margin by adding a percentage or specific amount on top of the buying rate and generate the user's tariff. Wholesalers usually work with many different vendors, so it is important to make sure that softswitch supports necessary bulk rate processing tools that simplify and partially automate the main tasks. The key tasks include a comparison of rate lists from multiple suppliers by quality and cost, exporting the final tariff and sending it to customers. Usually, each provider can offer at least three types of rate sheets, depending on the price and quality ratio. Some providers use tech prefixes to let their clients decide which route to follow. Here's an example:

- Grey route is selected when dialing the destination with the tech prefix 11.
- Standard route is selected when dialing the destination in the regular E.164 format without any tech prefix.
- Premium route is selected when dialing the destination with the tech prefix 111.

Another important part of the rate management process is pricing updates. Vendors change their prices pretty often and send notifications about effective and end dates of rate increases and decreases. Moreover route quality changes during that time, and has a direct correlation with the price (the better the quality, the greater the price). To schedule these changes, each rate update should be automatically imported into the system. If this affects customers, they must be notified by email. I've seen many providers struggling with these tasks by using a simple Excel sheet, but it takes a lot of time and causes a big headache. The person performing this work has to analyze thousands of rates. The huge amount of information gives rise to human-factor mistakes. Quite often, if transit providers do not have an automated way to perform rate management tasks, they prefer to work only with the key directions instead of the complete A-Z list.

Retail providers do not have the same problem because they use just one or two providers to terminate calls. Moreover, the buying rates rarely change, and even if they do, it barely affects the profit margin, so the tariff update process is often skipped to save time. Retail providers usually charge their clients based on usage or subscription. Usage charges are associated with service utilization, e.g. 1 cent a minute for a call to Argentina. Such services can have different billing increments (per second, 30 seconds, or a minute) and minimal usage fee. However, it has become more popular to offer multiple subscription plans for a user. These can be:

- Flat rate, which includes a specified amount of minutes to few destinations (e.g. $10/month for 200 minutes to the USA and Canada)
- Periodic fee for a specific product (DID, virtual PBX, or VoIP phone rent)
- Service bundle (e.g. $90/month for 150 TV channels, unlimited local calling and Internet connection)

In addition to this, retail providers may apply more advanced tariffs based on:

- One-time service, such as installation, activation or suspension
- Hour (peak and off-peak) or day (work day, weekend, and holiday)
- Time-limited promotional packages and discounts for long-term usage

- Extended billing schemes (set-up fee, daily charge, call split to intervals, minimal time payment interval, or grace time)

Extended billing schemes are especially important in the calling card business, which is often based on hidden charges. By reviewing call statistics and analyzing customer behavior, providers can adjust rates and add extra fees to maximize profit or increase voice traffic volume.

Routing

VoIP service providers usually interconnect with at least a few carriers and have multiple choices to terminate calls. To manage those choices wisely, they use a call routing feature that controls the voice traffic path to termination destinations. Softswitch can support a full array of call routing capabilities that help maximize the voice network performance.

Routing in the softswitch is managed through the routing plans that describe how calls to particular destinations should be sent. Providers can create unique routing rules for specific customers, set a global routing policy for a group of users, or isolate routing and make the high-quality providers available only for premium clients.

Routing by Price and Quality

The most common routing rule is LCR (least cost routing), which refers to the policy where calls are automatically sent to the most cost-effective path. However, quite often a low rate implies poor call quality, so routing only by price can result in low quality termination. In many cases, providers agree to deliver a certain quality level at an agreed rate. By lowering the quality, they risk losing trust and potential income from their clients. To keep the route quality at a consistent level, providers set routing rules that are based on key quality parameters, such as ASR, ACD, and PDD. In this case, the softswitch monitors the route quality in real time and re-evaluates the routing priority after a specific period.

Speaking of price and quality, it is important to apply the right proportion, depending on the target audience. You've already learned that as a solution, wholesalers offer three rate sheets that differ by price and quality. This helps set specific expectations for their clients in advance. If a provider operates in the retail market, it may be enough to have one tariff, but the quality and service stability should be at a sufficient level.

Additional Routing Options

If providers routed calls only by price or quality without considering other factors, they would not be very successful compared to their competitors. A few additional options can help you manage the routing more intelligently.

First is creating a dynamic routing plan to use in a combination of multiple parameters: profit, percentage, and priority. Profitability control allows you to "catch" the best opportunities and block those calls that are unprofitable. Setting a percentage allows distribution of the load by giving a specified traffic portion to different suppliers. This is quite common among those who operate a VoIP GSM termination business with multiple gateways in different locations. The termination self-cost is the same, so it is important that the traffic capacity is shared with the gateways evenly. Terminating all calls through one gateway would increase the risk of SIM card blocking. Priority is more of a human factor based on the know-how of the person who manages the routing (e.g. they know that certain providers always try to play various tricks by mixing traffic or doing FAS. Therefore, the priority is for a trusted provider, even though the price and quality are the same.) Each provider needs to find their own "secret" route management recipe that refers to the ongoing process of setting metrics, measuring the results, and adjusting the policy, based on what works best.

Another option is live monitoring of different parameters. This will be reviewed in more detail in the next subsection.

Optimum routing decisions have a high impact on your service level and can significantly improve your business image. Not all systems support the advanced tools described here. When you select the softswitch, make sure that it supports at least routing by using the key quality performance indicators: price, percentage, and priority. Anything that comes in addition to this can be considered as great value-added features.

Monitoring

Monitoring is tracking various characteristics of VoIP traffic. Wholesalers that exchange minutes with other carriers follow the live traffic and follow different quality metrics. Retail providers that serve end-users with SIP devices monitor network parameters and analyze SIP packets.

Alerts

Monitoring can be live or based on alerts (also called "alarms"). Live monitoring in wholesale business is done by a network operations center (NOC). A person, working in this position, makes needed changes by analyzing active calls and various statistics (if resources allow, NOCs work in multiple shifts by ensuring 24/7 availability).

Alerts are used to identify a change in pre-set parameters and to trigger an action. Providers use alerts to organize their operations wisely by automating most of the repetitive actions. Probably the simplest action is to notify the system administrator, client, or provider by email or SMS about the changes in monitored parameters. In this case, it will always require human interaction to perform the needed task. A more complex alert can program the system to take the action by itself without requiring a confirmation from the administrator. In this case, the system can change the routing or temporarily disable or block a client, supplier, or destination. It is especially useful in preventing client losses if someone has attacked their SIP device.

If a customer normally spends a maximum of $5 per day and the traffic unexpectedly increases to $10 in just a few minutes, an altering tool can automatically block such calls and notify both the client and the system administrator about a possible hacking attempt.

MOS (Mean Opinion Score) and R-value

Most problems in the VoIP business are related to poor voice quality; ensuring stable and reliable calling services is the top priority of each VoIP provider. Voice is very sensitive to network glitches, such as delay variation and packet loss. Because of this, it is important to analyze SIP packets and network parameters that influence the quality of VoIP calls. You are already familiar with the parameters, such as ACD, ASR, and PDD, widely used in wholesale VoIP. There is another way to measure quality by calculating the MOS and R-value. The MOS measures subjective call quality and scores in a range from 1 (unacceptable) to 5 (excellent). The R-value uses a mathematic formula that incorporates network latency, jitter, and packet loss, and grades on a scale of 1 (unintelligible) to 100 (very clear). Below is a table which explains how you can interpret each of those parameters:

R-value	MOS	Comment
90-100	4.2+	Excellent
80-90	4.0-4.3	Satisfied
70-80	3.6-4.0	Unsatisfied

Listening to the Call

The last way to monitor quality is to review call recordings or listen to the ongoing call in real time, directly from the GUI. This feature is sometimes called "spy real-time calls" and can also be used for legal interception. Listening to the call provides sound from the end-user's perspective. Before choosing which call to listen to, it is recommended to filter CDRs and find the worst call by jitter, latency, packet loss, MOS or other parameters.

Tips to Increase Call Quality

Here are some additional tips that will help to ensure a good call quality:

- Deploy the softswitch as close as possible to your clients
- Add multiple providers to the softswitch and choose automatic routing algorithms by quality
- Use monitoring and reporting tools to analyze VoIP traffic (e.g., VoIPMonitor or 5gVision)
- Choose only reliable equipment, preferably supporting echo cancellation
- Enable better quality codecs like G.711 or G.729; avoid using G723.1, G726, GSM and similar poor quality codecs
- Implement QoS awareness to prioritize packets
- Use additional tools (like Wireshark) to analyze network protocols and to pull jitter and network latency information from the packet capture

Using the above suggestions it is possible to deliver high quality sound which is almost the same or even better than legacy telephony.

Reporting

Monitoring is a real-time action that allows full control over the voice traffic at a given moment. In addition to this, most businesses need to analyze and compare information from different angles in a long-term perspective. To do that, various statistics, graphs, and reports are used.

Reporting refers to displaying VoIP traffic statistics and filtering them by different accounts (customer, reseller, originator, and terminator) or characteristics like destination, tech prefix, time, price, or quality. Statistics contain valuable information that helps to evaluate the health of a business and identify upcoming problems. Proper reports allow fast decisions to be made in order to improve business performance. A commercial team keeps track of financial reports, such as income, profit margin, and commissions for sales managers. A technical team evaluates traffic volume, system load, and quality performance reports. Reporting can be done in real time (like monitoring), periodically (hourly, daily, weekly, or monthly), or based on events (alerts).

If a softswitch misses some statistics, an additional reporting tool can be used in parallel with the main system. In this case, you just need to export CDRs from the GUI or retrieve them directly from the database.

Account Levels

The highest account level in all systems is the administrator (in short, admin). Admin has full, privileged control over all aspects of the system. A majority of daily operations and the regular system setup are done through the web-based panel (GUI). This ensures easy administration and enables access for other internal and external system users.

Internal users are employees of the company: accountants, sales managers, engineers, and network administrators. To ensure business and data confidentiality, a system administrator has to arrange multiple access levels with certain rights for each position. Managing permissions is also important for security reasons because an employee who does not know how to operate the system properly may do some misconfigurations that can cause serious troubles. To avoid this, each account can perform only those tasks that are musts for that role. For example, sales managers should be able to add and change tariffs, rates, profit margins, payments, and review various statistics. Engineers require functionality that helps them troubleshoot problems and implement needed repairs. Accountants must have access to manage payments and invoices.

External users are everyone who uses the system, but who does not work in the company. This is usually customers, resellers, and providers. Similar to the internal control, each group of external users has unique access level to the softswitch interface.

Customer interface is usually provided for retail users if the service provider has an online presence. The user panel is like a self-care portal, which allows performance of most of the following tasks:

- Sign up and log in
- Check credit or balance
- Choose the service or subscription plan
- Purchase DIDs and forward them to the needed number
- Review CDRs, payments, call statistics and reports
- Top up the balance
- Pay the invoices

Some providers deliver their services through resellers and give them a portal to manage their main activities. In addition to the simple user functions, resellers can:

- Add new prepaid and postpaid users
- Manage different services (calling cards, callback, DIDs, and flat rate subscriptions)

- Have a multi-level (hierarchical) structure that allows the creation of subresellers with different rights (enabling resellers to expand their sales channels by finding service distributors or agents)
- Customize the portal with own brand (upload logo, add the company name, or change colors)
- Review revenue reports
- Create own tariffs and manage profit margins

With regards to the last option, there are two ways for a provider to split the profit with resellers: provide commissions or allow them to set their own margins. Working on commissions is a preferable method by those resellers who are not tech-savvy and do not want to manage additional operations. However, more advanced sales partners prefer to have more control and need to feel a stronger sense that they are the business owners. In this case, providers can allow resellers to create their own tariff plan and set their own margins. Having such permission ensures more opportunities to make a higher profit for resellers and gives them the freedom to make their own decisions. If a reseller wants to have even more control, there is a feature called "switch partitioning." Partitioning allows the management of their own providers and routing plans. It is a good alternative for those whose business is growing, but who cannot afford their own softswitch yet.

The last account level is the provider. The provider interface is a great tool to review billing information from the perspective of different systems. Carriers can check whether the billing matches their data and, if not, can manage disputes accordingly. Moreover, such a panel is useful for those providers that do not have their own billing system, but who would like to keep at least minimal control on the voice traffic they terminate. It is especially relevant for individuals or organizations that provide call termination through VoIP GSM gateways without a softswitch.

API (Application Programming Interface)

Most VoIP providers use multiple softwares to manage their operations (website, accounting software, CRM, ERP, or mobile dialers) in parallel with the softswitch. Quite often there is a need for programs to interact with each other by importing or exporting information or triggering a certain action after specific conditions are met. In such cases, providers seek an integration of all elements and automation of the overall system by using API.

API is a set of routines to manage a system to integrate with another piece of software. This method allows you to insert or retrieve data from the database securely and to control the process of a specific action. All the processes run in the background and are invisible to end-users. One of the most common examples is an online purchase. Online stores allow users to enter credit card information, which is then stored and sent to a remote application using the API to verify whether the information that the user entered is correct.

When choosing a softswitch, find out if it has an open API that is usually published by the vendor and can be used freely for integration with third party software. API is one of the most popular integration methods, but systems may support additional ways, such as MySQL scripts or a special development kit.

CDR Reconciliation

CDR is the main source of information about the call, so the accuracy of this data has a direct relation to the service provider's revenue. There are many cases when CDRs from different sources do not match, and providers need to run the CDR reconciliation process, which finds discrepancies between the carriers' amounts. Differences can show up for various reasons, including the loss of CDRs in the carriers' internal systems, call duration difference, mismatch of called numbers, rating difference, and incorrect rounding.

CDR reconciliation is more relevant for carriers that exchange traffic in the wholesale market. The automated reconciliation process allows a reduction of time and operational efforts on disputes between providers due to mismatched CDRs. Usually, the outcome of CDR reconciliation is a financial figure that is negotiated and posted as an adjustment into the invoice. Studies show that in wholesale VoIP around 3% of revenue is leaked because of different types of losses, due to mismatches between different CDR sources.

Auto Dialer

An auto dialer is a feature that enables a voice broadcasting service. It allows you to send phone calls with a recorded message to a list of call recipients (members, subscribers, employees, or customers) automatically.

The auto dialer can be passive or interactive. Passive will broadcast telephone messages to call recipients and allows you to review the statistics afterwards. Interactive allows the call recipient to listen to the recorded message and interact with the system by pressing keys on the phone keypad.

In this case, a system can play a promotional message like "Dear customer, today we have a special price for call termination to Cuba. If you are interested in this destination, please press 2 to learn more." After a person presses the key, they will be connected to the sales representative to place an order or to receive more information regarding played messages.

The target audience for voice broadcasting solutions can be anyone who would like to reach their clients. But overall, it is mostly used for political campaigns, interactive polls or surveys, reminders (appointments or payments), attracting more leads, or delivering information to a long list of recipients quickly.

Creating a successful broadcasting campaign consists of the following steps:

- Writing a message that is powerfully persuasive
- Recording the message using professional voice narrators
- Uploading a list of numbers of people who should receive the message
- Launching the campaign at a specific time
- Reviewing the reports of a launched campaign and analyzing the results

It is important to note that auto dialers are prohibited in some countries, so contact your local authorities before using this feature.

Predictive Dialer

A predictive dialer allows you to dial a list of numbers; if the call recipient picks up the phone, they are connected to the next available representative or agent. Different from an auto dialer, a predictive dialer is programmed to predict when a call recipient is available to pick up a phone. Only a low percentage of answered calls are abandoned. A predictive dialer monitors the availability of agents, tracks the average call duration, and measures other factors that help determine the most efficient outbound dialing frequency. It is widely used by call centers and allows agents to work more efficiently by minimizing the time they would otherwise spend waiting between conversations. (Probably the most popular open source solution used as an outbound predictive dialer is Vicidial.)

Call Shop

Call shop is a web-based or desktop system for monitoring and controlling calls in phone booths and managing billing and invoicing processes. Usually, call shop cashiers and administrators are not tech-savvy, so one of the key factors for this interface is simplicity.

The main functions that should be supported in call shop systems are:

- **Prepaid and postpaid accounts.** If a user pays in advance for a call, a cashier enters only this amount and, once the call is completed, returns the change to the user. In postpaid case, there should be a credit limit for each call session.

- **Call control.** A cashier should be able to switch booths on and off anytime and, if needed, disconnect calls in progress.
- **Invoice generator.** After the call is completed, a cashier should be able to generate an invoice or receipt.

Also, it is important to note that customers feel more comfortable when they see the balance during the call session. Because of this, call shop owners must ensure displays in call booths that will show such information.

Calling Cards

Calling card features enable you to manage card groups, generate PIN codes in bulk, set various fees, and apply additional "tricks" to optimize your revenue.

Providers can sell calling cards directly, through distributors, or resellers. The more parties which are involved in the distribution chain, the lower the profit margin for the provider. If you plan to work with sales and distribution partners, you will need to provide a user-friendly web portal where they can keep track of calling cards sales and review the revenue or commissions. Most providers allow both prepaid and postpaid accounts for their partners and minimize their risk by deducting the balance or charging them only when a card is used the first time (also called being "activated" or "scratched").

Most calling cards inform a caller about the terms and conditions in the small print. It looks similar to an advertisement. A user sees a great offer in large print (in this case—balance of the calling card and a number of minutes).

However, this "deal" is provided under certain conditions (e.g. balance of the calling card should be consumed in one call to a particular destination). The trick is that, statistically, only a low percentage of card users can stick to those terms. In this case providers may apply different charges:

- **Activation fee.** Deducted when a user activates a card the first time.
- **Connection fee.** Applied each time a user makes a new call.
- **Disconnection fee.** Deducted once a call is disconnected. This fee can be fixed or proportional to the call length or price.
- **Ghost percent.** The message, played by the IVR, which announces a higher balance or number of minutes than a user actually has. If the card value is $5 and the ghost percentage is 100%, the message will announce that the user has $10 in the balance. It usually surprises and excites users, thinking the provider has made a mistake, and they will be able to take advantage of it.
- **Maintenance fee.** Deducted from a user by a certain frequency: each day, week, or month.
- **Multiple use fee.** Applied as a "penalty" if the user consumes a card not at once, but in multiple times.
- **Peak and off-peak rates.** Based on the time of day (working hours, lunch time, or non-working hours). Busier times usually have a higher rate.
- **Rounding.** Charging a user based on increments, such as per-second, per-30 seconds, per-minute, or longer.
- **Step billing.** The application of a different rate depend-

ing on the period of the call. As an example, you can apply 4 cents for the first minute, 5 cents for the second minute and 6 cents for further minutes. If a user talks for 3 minutes, the charge will be 15 cents.

- **Surcharge**. A fixed fee or a percentage that is added to the regular rate. If the rate is 4 cents per minute and surcharge is 10%, the card user will be charged 4.4 cents per minute.

Callback

The callback feature allows the management of callback services. The most popular method to trigger a callback is by calling a DID number. But this is not the only one. There are a few additional ways to initiate the callback service:

- Sending an email in a special format
- Sending an SMS in a special format
- Entering the source and destination numbers in the user interface
- Using an API request (for integrations with the third party applications)

The call path in callback begins in the mobile operator's network. Then it goes through the DID provider until it reaches your system. Totally there are at least two providers between you and your client, which means that you do not have full control of the call connection. Therefore, before using a callback, there are few important aspects to consider.

Firstly, make sure that your DID provider sends the correct Caller ID information. If not, your system will be unable to determine the original caller. Secondly, it is not guaranteed that a mobile operator and DID provider will send the DTMF (dual-tone multi-frequency signaling) tones properly. This means that the caller may not be connected to the call recipient, even though the destination number was dialed after receiving the dial tone. You can choose or change the DID provider, but you will not be able to do the same with the mobile operator. So test your callback service with all major mobile operators in your market before launching this service.

The last thing to note is that if you apply authorization by PIN in a callback service, and the caller is not authorized for some reason, you will be the one who has to cover the calling expenses. This happens because of the callback logic: system calls back any caller who knows the callback DID, but it allows dialing to the destination only for those who enter the right PIN code and are authorized. The sad truth is that you'll have to pay for all your clients who enter the incorrect PIN a few times in a row.

Also, you may receive calls to your callback DID number from people who are aware of your service and are hoping to guess the PIN code or who are trying to harm you (e.g. if they are your direct competitors). To avoid this, use authorization by ANI (pinless). In this case, the system will call back only if the caller ID matches the records in its database.

Dial Plan

The dial plan is an integral part of a call process and is widely used in all communication systems, including phones, PBXs, and softswitches. Dial plans can be configured by using a specific syntax where each symbol has a certain function or by using a web interface. The main role of a dial plan is to instruct a device or system how to carry out and route calls by performing the following functions:

- Addressing the destination or endpoint (phones, softphones, voicemail systems, and auto attendants)
- Managing calling roles and privileges (some phones can reach only internal numbers without the ability to call outside)
- Adjusting the number. As an example, the dialing format in the USA to reach a destination in the UK might be 0114412345678, where 011 represents the international dialing number. Whereas in Spain, the format could look like this 004412345678. In this case, "00" stands for the code before an international number. Such formats and rules should be defined in a phone, PBX, or softswitch to make sure all calls are passed successfully.

Usually, there are some default dial plans in softswitches. They come with a pre-defined logical sequence. Examples of such dial plans are: authorization by PIN, authorization by ANI, auto-attendant, calling card, and callback.

DID Management

DID management is a Class 5 feature, more relevant for retail business. DIDs can be obtained from international DID providers (some examples will be provided in the next chapter), local carriers, or allocated by the regulation authorities.

DIDs are usually managed in an E.164 format and can be marked as private or global. Private DIDs are assigned to a subscriber or a company and are used as virtual numbers for incoming calls. Global DIDs are usually assigned to a specific dial plan and used as an access number.

When choosing a Class 5 softswitch, it is important to pay attention to the possibility of managing the DID billing. It should have similar capabilities to regular tariffs and enable free DIDs (e.g. as a value-added service if another, main product has been purchased or if a long-time contract has been signed), monthly subscriptions, or advanced billing schemes, such as peak and off-peak rates, a connection fee, or a simple per-minute charge. If DIDs are used as an access number, make sure that your softswitch supports Leg A and Leg B billing separately, as sometimes calls through the access numbers should be charged extra (e.g. when using a toll-free number).

Nowadays, most VoIP providers have an online business, so it is important to understand the DID management process if you sell your services through a website. Consider providing a self-care portal for your clients where they can choose DID numbers from the list and purchase them by using an online payment gateway. Once DIDs are obtained, users should be able to route them to an IP address or hostname, forward calls to the PSTN or SIP number, and review statistics on DID usage. Usually, the bigger DID suppliers offer a possibility to manage DIDs through an API. It allows automation of the buying and DID assignment processes.

Number Portability

Number portability is important for managing the routing if provider charges are based on on-net and off-net calls. Another use is the technical procedure of number porting.

In local interconnections, there are usually two types of pricing: one for call termination within the provider's network (on-net) and another for call transit outside the network (off-net). Calls within the provider's network are usually cheaper than a transit of calls. The question is how to determine whether the dialed number belongs to an interconnected provider, if the prefix does not reveal anything about itself. In every country that allows portability, there is a national or private number portability clearinghouse that contains ported numbers. It shows which provider the number belongs to at a specific time.

To optimize the revenue, the softswitch interacts with this remote database and routes calls through the provider, to which the number is assigned at that moment. If the number is not found in the database, the softswitch will route calls based on the prefix. However, the need of this feature depends on the percentage of calls that are not routed through the optimal path. If there is just a small percentage of such calls, the investment of including number portability add-on and automation may not pay back.

Another important aspect is the technical number porting procedure. Such procedures are usually issued and controlled by regulation authorities. Those providers that port a lot of numbers are seeking full porting automation. But for smaller VoIP providers, it is not as relevant. Either they rarely have such requests or number porting is done through the third parties.

Multi-tenant PBX

All Class 5 features, such as voicemail, call waiting, follow-me, queuing, and hunt groups are supported by most PBX systems. This gives the service provider a few choices: to serve clients with existing PBXs, to install and maintain on-premises PBXs, to set up a virtual PBX that resides in the service provider's datacenter, or to select a softswitch with multi-tenant PBX capabilities.

The last option is preferred by most providers since it allows using a single system to deliver services to multiple clients (also called "tenants"). Multi-tenant PBX gives the following benefits:

- **Cost savings.** There is no need to invest in additionnal hardware to enable PBX for a new client.
- **Easy scalability.** New accounts can be activated through the self-care portal. The system administrator can forget about installation and setup of the new hardware.
- **Simple management.** It is much simpler to maintain one system than tens or hundreds of servers for each customer.

However, it is not very easy to find an all-in-one softswitch with multi-tenant PBX functionality that would be affordable and suitable for all PBX-related services. As a result, some providers end up using two separate systems—one to deliver the PBX function and another to manage routing and billing.

Transcoding

Sometimes clients and suppliers use different protocols and codecs. If they do not match, a softswitch should convert a voice stream from one format to another. The process of matching end-user device and provider capabilities is called transcoding (for codecs) and protocol conversion (for protocols).

Transcoding allows two devices or systems without common codecs to communicate with each other. The most frequent conversion is between G.711 and G.729 codecs. It is especially common in wholesale VoIP for establishing interconnections between different providers. The transcoding is built upon complicated algorithmic computations and requires significant system processing resources. As a result, each time the format of a call needs to be converted, the load in the softswitch increases. Wholesale business profitability is based on volume, so capacity limits are a big challenge for transit providers. They try to prioritize calls that are originated and terminated in the same format. If the part of transcoded calls is relatively small, you can use a softswitch to perform this function. But if the percentage becomes really high, it is recommended that you pass the conversion function to third party systems, such as transcoding cards or gateways. They allow you to optimize the system performance.

In addition to codec conversion, a softswitch can manage connections of two streams with incompatible signaling capability. A VoIP protocol conversion, such as SIP, H323, and IAX2, is usually done in the softswitch. But when there is an interconnection between two different signaling technologies like SIP and SS7, providers deploy a gateway or BRI/PRI card to do the job.

SBC (Session Border Controller)

One of the important security elements that hides and protects VoIP network devices from various public threats is an SBC. It can come as a standalone solution or as a partial softswitch function. An SBC acts as a gateway between the enterprise and provider networks (accordingly there are two types of SBCs—one for enterprises and one for service providers). It allows only authorized calls (sessions) to pass through the connection point (border).

Most users initiate calls from devices that are behind some border which stands between local and public networks. This border can be a gateway, router, firewall, network adapter, or some other system that connects two types of networks and performs network address translation (NAT). NAT refers to the Internet standard that allows users to communicate inside and outside the local network by setting two sets of IP addresses. One set is for internal traffic, another is for external traffic, such us public Internet browsing or registering a VoIP device to the SIP server.

However, when calls are being sent through the network, a simple router or gateway cannot guarantee proper NAT translation. As a result, users may experience call quality issues or face problems with SIP device registration. In this case, an SBC ensures the seamless communication between network devices through the use of a variety of techniques, such as NAT traversal, SIP message, header manipulation, and protocol translations. Moreover, an SBC prioritizes the voice traffic flow (especially emergency calls and lawful interception) and ensures the quality of service.

From the functions above, a session border controller may seem similar to a softswitch. However, the main role of SBC is to protect security threats in an IP connection, whereas softswitch is meant to be used to connect calls by applying various routing algorithms.

Comparison Elements

Now, as you are already familiar with the main and additional softswitch functions, it's time to select the right one for your business. There are many different ways to compare vendors and their systems. In this section you'll find the most important characteristics to consider before choosing the softswitch.

Softswitch Type

You've already learned about different types of PBXs: hardware vs. software, on-premises vs. hosted, legacy vs. IP-based, and open source vs. proprietary. All of those differences are also applicable for softswitches.

When you start comparing different switching systems, you'll discover that a majority of them are software-based. It is the most preferable option as you can run the software from anywhere. It can be installed on your computer at home; you can place it in the server rack, or you can choose to set it up in a datacenter located in another continent. Moreover, it is much easier to set up redundancy, make a backup, or migrate data and license to the new server if the main one is down.

If you do not have a skilled technical team or personal experience with managing hardware, choose a cloud-based softswitch. This will enable you to manage the system through the user-friendly web panel, and your service provider will take care of all the hardware-related issues. On the other hand, if you serve regional clients and the equipment of the softswitch vendor is too far away from your clients, consider a premise-based solution. A great distance can cause latency issues, so it is better to have the softswitch as close as possible to your clients.

Another important consideration is whether to choose an open source or a proprietary softswitch. On this subject, people usually have one of the following stereotypes in their mind. Open source users are often perceived as companies or individuals without sufficient budget, who are accepting the risks relating to the product stability, performance, and security. Proprietary software users are viewed as enterprises with enough cash flow in their accounts, who are therefore looking for more robust and stable commercial products which come with a support contract, provided by a credible and well-known organization. Even though these choices make sense from the financial perspective, there are still other reasons why users choose a specific software type.

Here are two reasons to select an open source system. If your business model requires something unique that is not available with proprietary solutions (or those functions cost too much), use the softswitch with an open architecture. It will be easy to customize and implement needed changes. Another motive for choosing an open source system is if you have a strong technical background or an experienced team of software engineers. In this case, managing an open source system will not cause too much trouble.

The last choice in softswitch types, which does not exist in PBXs and is applicable only for softswitches, is the class. It is quite easy to determine which class to choose. If you plan to manage a wholesale business, you will need a Class 4. If you want to work with retail clients, you'll require a Class 5 softswitch. In the end, if you are not sure what business you will manage and want to have the freedom to decide, begin with a mixed-type softswitch that incorporates functions for both classes.

Technical Specifications

The second comparison element is technical specifications. In this subsection you will learn more about different softswitch characteristics that you can easily compare while evaluating different vendors.

Hardware, Network and Virtualization

If you plan to use a software-based switch, you will have to prepare a server. Each software vendor has specific hardware requirements for their softswitch.

Normally, hardware requirements consist of four main elements: processor (CPU), memory (RAM), hard disk (HDD) and operating system (OS). In addition to these, you may receive general advice on hardware brand. If not, it is always better to stick to the reputable name. Server specifications may vary depending on the performance you expect; the higher the voice traffic volume you want to transmit, the better the server you'll need. A softswitch provider may give you different hardware recommendations based on the call capacity, or else set the minimal requirements to run the system. Moreover, if a softswitch is not multi-functional and works in a distributed environment, each of the components (database, billing server, web server, and load balancer) will have slightly different requirements.

Another important element is network specifications. You may need to open necessary ports to access your server and set up a public static IP address, preferably with a broadband Internet connection, so that your system will be reachable by clients and providers.

The last thing to mention is virtualization—the technology that creates a virtual version of a server, operating system or network resources. It is quite popular—especially among startups—as it helps to reduce the initial investment on hardware and simplifies software management. Virtual machines allow you to perform a quick recovery by taking a configuration snapshot, if needed. However, it is important to note that software installed on a virtual machine may run slightly slower than software running on a native host (dedicated server). Moreover, not all softswitches can run smoothly at a higher load on such machines. Always consult with your vendor before choosing this path.

Architecture

Architecture refers to the softswitch components in a distributed environment. You've already learned that components can be unified in a single multi-functional softswitch or installed on several servers in a cluster configuration. The first solution has a lower price and is easier to manage while the second solution achieves better performance but requires a higher investment. As an example, a five-server solution may consist of:

- A RADIUS billing engine which performs AAA, rating, routing, and service provisioning
- Two database servers with MySQL replication (data duplication for backup)
- Web server for system management
- RTP server for media applications (voice stream, call conferencing, IVR, and unified messaging)

You need to predict your business growth before choosing the softswitch architecture. For most startup providers it is recommended that you start with a single server solution or two-server redundant solution.

Performance

The key performance characteristics are concurrent calls and calls per second (CPS). But sometimes softswitch vendors provide additional measures, such as busy hour call attempts (BHCA), busy hour call completion (BHCC), the number of registered devices or subscribers, and monthly traffic in minutes.

You may learn that the call capacity of some softswitches may differ significantly compared with others. The highest performance level can be achieved by carrier-grade Class 4 softswitches which run in a cluster environment. Such implementations are designed for short-call-duration traffic and high peak loads. Here are the main factors that influence softswitch performance:

- **Hardware.** The better the server you use, the more calls you'll be able to handle.
- **Call length.** Shorter calls (so-called "dialer" or call cen-

ter traffic) place a bigger load on the softswitch because most resources are used to establish a new call.

- **Transcoding.** It takes on average two times more resources than a regular call. If possible, it is recommended to use a transcoding card or external gateway to perform codec conversion.

- **Call structure.** Call consists of signaling and media. In a distributed environment, signaling and media streams can be separated. If you pass only signaling through the server, the softswitch may handle ten times more calls compared to transmitting calls with media. When reviewing the performance results in softswitch specifications, pay attention if the results are shown with media pass-through (RTP stream is passed through the server) or direct media (only signaling is passed through the server).

- **Processes, running at the same time.** Normally there are additional processes that run in the softswitch while it handles calls. You may need to generate invoices, review statistics, or import rates. If everything runs on one server, it will influence the processor load, and softswitch performance will decrease.

- **Business model.** The simplest business model is a wholesale transit, where the system just passes calls through from one point to another. However, if you provide calling card services, softswitch will need to play an IVR before each call, authorize the user, and handle both Leg A and Leg B. As a result, the call performance in a calling card business will be much lower compared to the wholesale transit.

- **Switching engine.** Many softswitches are built on open

source platforms, such as Asterisk, FreeSwitch, or OpenSIPs. Asterisk-based solutions provide a wide set of Class 5 functions, but do not work well at a high load. Solutions based on FreeSwitch and OpenSIPs (Kamailio), however, run much faster, but do not include the advanced features you find in Asterisk. Moreover, vendors can use their own-built switching application if none of the open source platforms matches the product vision.

- **Database type.** Each time a call is established, a record in a database is created, so it may become a bottleneck when there are many SIP servers that handle high voice traffic volume. Similar to the switching engine, vendors may use an open source database (MySQL or PostgreSQL), which reduces the softswitch cost but has a lower performance or a commercial database (Oracle) that runs faster but comes at a higher price.

- **Distributed components.** When you compare different softswitches, take into consideration whether they come as a single server solution or as a cluster. In a distributed environment, softswitch elements run independently from each other on separate servers. Running only one specific process on a single machine allows an increase in the overall system performance. It is quite common to separate the following elements (if such an option is allowed by a softswitch supplier): database, GUI, billing (RADIUS), media (RTP), SIP signaling and load balancing.

Each softswitch has its performance limits under certain conditions. If the load reaches a critical level (100% or more), stability issues can arise. A softswitch may then start to delay calls or to reject new connection attempts. In such cases, you may need to restart the system, causing downtime and service outage for your clients. To avoid this, be aware of softswitch limits and find out if they have some protection mechanism that helps prevent the system from overload. This knowledge will help you to plan your business growth.

Scalability

Most VoIP beginners think only about calls when reviewing softswitch performance. However, the actual performance bottleneck may appear in other elements, such as the GUI or the database. If your client base and traffic volumes are growing, you will need to improve the performance indicators (upgrade hardware, reduce transcoding, or separate signaling from media), or scale the system. Scalability describes how easily you can expand the system if you are reaching the limits.

Most softswitches are scalable, allowing expansion of the call capacity by adding more servers and forming a multi-server cluster. To create a thorough traffic growth plan, there are a few important questions that should be clarified with the softswitch vendor in advance.

- Is load balancing available? If not, you may need to control the balancing to multiple SIP servers manually (which is not efficient).
- What implementations are available, and which of them are recommended for your growth plan? If possible, ask for examples of topologies that are already implemented for other clients and are well tested.

- What are the costs associated with new implementation or adding an extra element?
- How much time does it take to change the primary implementation or to add more servers to the cluster?
- Will there be downtime of services during the expansion? If so, how long?

It is essential to make sure that the softswitch provider will be capable to meet your growing voice traffic demands and that the expansion will be seamless for you and your clients.

Redundancy

In the VoIP business there is always a risk of hardware or software failure, so you must use a proper deployment to ensure high service availability. You can save data and configurations by organizing manual or scheduled backups or by implementing an automated failover (redundancy).

There are different names for redundancy: high availability, failover clustering, disaster recovery, hot standby, and dual-plane architecture. In general, it ensures the maximum uptime of the softswitch (you may have seen some datacenters promoting their server availability as "five-nines" or 99.999%) and eliminates a single point of failure. It is achieved by duplicating data and other critical system components. In some countries, regulation authorities treat redundancy as an obligatory deployment for all telecom service providers. It is especially important for the emergency calling feature. It is your responsibility as a service provider to ensure that your clients can call an ambulance or the police in case of emergency, despite the fact that your main server is down.

The minimal failover solution requires two servers. In this case, another identical server (also called a "backup" or "slave") is deployed which is mirrored and synchronized with the main one (also called a "master"). It can be an active-passive configuration where only one server accepts calls, while the second one works as a backup and takes control of calls automatically if the main one is down.

Another type of configuration is an active-active, where both servers can handle calls at the same time, but you may lose some traffic if the main server is down and you have more calls than one server can handle.

In most cases, redundancy is configured in the same location or datacenter. However, some providers are afraid of datacenter failure and consider making redundancy between different countries or even continents. However, disaster recovery between two geographically distant datacenters is a high-level feature, and not all softswitch vendors can ensure this.

Integration

Most proprietary softswitches come with a "static" GUI, which cannot be modified. However, your clients may prefer to have a more user-friendly interface or a customized panel with only certain features available. In addition, you may need to link the softswitch together with third party software, such as your website, mobile dialers, payment processors, ERP, CRM, or other systems so they will act as a coordinated whole. In all of those cases, it is very convenient to use an API, which is the most popular method of integration.

An API allows performance of different operations without involving the GUI, and you don't need to know the programming code of the system. However, you do need to know in advance what applications and platforms must be integrated with your softswitch and what information you need to place or retrieve. Once you are aware of this, make sure that your softswitch vendor has the required API for it.

In addition to the API, integration can be done by using scripts in the database. However, this is not convenient, because you must have a description of the database architecture and understand how to operate it. Moreover, this method is not as secure as using an API—if you change some important database fields, you may ruin the system. So in this case, always contact your softswitch vendor before you make any changes in the database.

The last thing to note is that two machines are recommended when you undergo any integration. The main one is used for the production environment, the second for testing the integrations and modifications. Only when everything has been well-tested on the "development" machine should you use the integration scripts in the production server.

Security

The third comparison element is security. This subsection provides a wider overview of security threats in a VoIP environment.

Together with the development of VoIP technology and open source systems (such as Asterisk or FreeSwitch), provider-level infrastructure has become available for almost anyone. This has also led to an increase in VoIP attacks, as those with more technical telecommunications knowledge have started to specialize in IP telephony in order to to get into the illegal use of the worldwide VoIP market. Today, malicious users are constantly trying to identify existing VoIP weaknesses and are using different tools and techniques to get into VoIP systems.

The first type of hacking includes everything that causes harm—denial of service (DoS), interception and modification, eavesdropping, and Caller ID spoofing. One of the best-known threats on the Internet is the denial of service attack. DoS can be achieved in different ways: user call flooding, registration or media session hijacking, request flooding or looping, and malformed protocol messages. Sometimes DoS attacks have cascading impacts; disruption of service may not be the main objective. It can also mask the intent to identify vulnerabilities and collect information that can be used later to benefit in other ways.

Another attack during which a hacker gets full access to the signaling between two or more endpoints is called "interception and modification." It allows an attacker to have complete control of calls and the way they are being connected and routed. Sometimes interception may be used not take over the control of the system, but to secretly listen to a conversation between two or more people. This process is called "eavesdropping" and can be achieved by intercepting and logging traffic with the packet or network analyzers (e.g. Wireshark).

The last type is called "Caller ID spoofing," which causes the telephone network to display a different phone number (Caller ID) instead of the original, from which the call was placed. The caller's original phone number may be 4412345678, but a call receiver sees the number 3312345678 on the phone screen.

The spoofing method can be used for many different reasons and sometimes it may bear no relation to hacking. Full disclosure: I tried to use Caller ID spoofing once when I had an unpleasant situation with a car mechanic who had been keeping my car for weeks and wouldn't pick up my calls. Luckily, I had his girlfriend's number, so I tried to use that Caller ID instead of my personal number. The result was great! All of my calls were suddenly picked up, and I had a chance to have a really weird conversation. However, I never told the auto mechanic how I did this, so I guess he still thinks that somehow I visited his girlfriend whenever I wanted to reach him.

Another type is "toll fraud," which is the illegal use of VoIP traffic without the intention of paying for it. It is the most popular among attackers. By combining different tools, hackers hijack a VoIP system and place unauthorized calls to premium numbers or expensive international destinations like Cuba, Sierra Leone, or Somalia. Illegal voice traffic will not stop until the hijacked line is blocked. In this way attackers can make a lot of money in a short time; the self-cost of calling is equal to zero while profit margins for call termination to those numbers are very high.

In the picture below you'll see a simplified VoIP network with three points that are vulnerable to exploitation: end-user device, IP PBX, and softswitch.

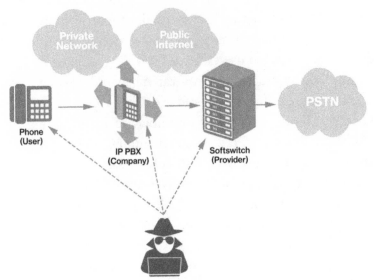

Figure 4.01. Points that are vulnerable to exploitation.

Usually end-user devices and communication systems become susceptible to security breaches due to:

- Weak password protection
- Limited policy enforcement (allowing redirects, transfers, and forwards to international numbers without proper authorization)
- Weak authentication (accepting calls from anywhere and routing them to the service provider with limited or no authentication)
- Vulnerabilities in open source VoIP solutions

Those who are planning to launch a VoIP business for the first time should note that VoIP attacks are often started just a few minutes after a VoIP server is connected to the public Internet. Another interesting fact is that most attacks take place outside of working hours, so that no technical personnel are likely to be present to monitor the system.

How can hackers be so knowledgable and speedy? Well, they know good tools and techniques that help them to identify the IP addresses of SIP servers and find their vulnerabilities quickly. Those tools are used to scan the range of IP addresses and identify target IPs or the entire subnetwork that uses SIP.

If SIP ports are open, hackers run a brute-force password-cracking attack. One of the most common tools for IP scanning and brute-force is SIPVicious, originally designed for SIP systems auditing. There are many cases where people do not pay enough attention to the strength of their password. After many failed log-in attempts, SIPVicious can eventually find the right combination to allow them to get into the SIP server.

The level of damage done by a VoIP hacker usually depends on the calling volume capabilities reflected in a user's balance (prepaid scenario) or credit limit allowed for this user (postpaid scenario). If a service provider has a client on a prepaid basis, the most that a hacker can do is to use the client's complete balance. Most probably this will not cause big disputes.

However, the situation changes dramatically if a hacker gets into a PBX or softswitch that is used for interconnections with other carriers with big credit limits. In this case, a hacker can cause the loss of tens or hundreds of thousands of dollars for a service provider by routing traffic to expensive destinations or premium numbers. Hackers make money by selling such traffic to other providers (who may not have any clue they are using illegally-routed traffic) or by getting generated income from premium number owners.

Hackers are rarely caught and punished, leaving the victim with the bill. Nevertheless, each situation is different, and many are resolved during disputes, and some of the losses can be cut by the supplier. However, it is far better to spend some time increasing the VoIP network security before anything happens. (You'll find the list of security suggestions in the next subsection.)

Evaluation

Evaluation is the most important aspect of comparison as it involves the real experience of the product and the customer service that comes along with it. There are usually a few ways to evaluate the softswitch: online demo; download of a free, limited, or trial version; and a pilot project.

First, it is recommended you play around with the system by using an online demo. You may be able to access the administrator and user interfaces by yourself or schedule a demonstration with the softswitch vendor representative, who will guide you through the main functions. Reviewing the interface will give you an idea about the look and feel of the product. Moreover, you can evaluate how user-friendly and intuitive the interface is. Usually a demonstration takes about an hour, and afterwards you can decide whether you'd like to test the product further.

If your experience with an online demo is positive and it has met most of your expectations, you can continue testing by downloading a free version or by requesting a trial or pilot project. Free versions and trials usually do not cost anything, but they either have some restriction on features or are limited by time (a month or two).

A pilot project is a paid option, but is a more involved experience than the other options; the software vendor puts in more effort by helping you configure a successful proto-type of your business model. A pilot project usually comes with some initial training, technical support, and a pre-defined goal that you agree with the softswitch vendor prior to running the project. This helps you focus on the mutual goal and evaluate not only the product, but also the team be-hind the softswitch. During the pilot project you can set up your "beta" business model and test it with a few close clients or friends. If you achieve your goals by the end of the pilot project, and you are happy with the product and technical support that comes along with it, you can invest in the product with confidence.

It is very important to set clear goals for testing. Do not use any generic terms, like "see how it works," "launch the service," or "evaluate all functions." The more specific you are about your expectations, the easier it will be to achieve them. In addition to the testing goals, there are some important elements you need to address. Below are the key recommendations that you should take into consideration before making a final decision on a softswitch.

Feature Set

You've already learned about various functions that may be found in a softswitch. The majority of them are feature-rich, but do not base your decision only on this. A more important aspect is to understand which features are a "must-have" for your business and which are a "nice to have."

Your priority is to evaluate the key functions and to make sure they are working properly and are easy to navigate and manage. If you are missing some functions, ask your supplier if there is any kind of workaround to achieve what you want; if not, can such a feature be implemented. Most people expect that the supplier will be willing to implement their requested functions for free because they are features that will be useful for everyone and will make all their users happy. However, the truth is that those softswitch vendors with a big client base already receive hundreds of requests each month and need to prioritize them. If your requested function meets the supplier's vision, there's a chance it will be implemented in the same or next version. But if it has a lower priority, the only option to get a feature is by paying for custom development or by negotiating to get this service free if you purchase the softswitch solution.

Usability

How easy is it to operate and manage the system—for you, your employees, clients, and resellers? A simple and intuitive interface allows you to get used to the system faster with minimal time consumption and less training from the experts. Simplicity is one of the key factors in the retail segment where end-users require the highest comfort level. In the competitive VoIP market, quality and price ratio are equal factors, so clients prefer to choose the provider that will most make their life easy.

Usability depends on the front-end (GUI) layout, the menu structure, and the previous experience you have had with other systems. It's no secret that when someone is used to one system, it is hard to switch to something else. Because of this, system administration should be simple enough to ensure a swift new user adoption. Normally within a week or so, you should be able to determine if a system is convenient for your business environment.

Automation

Each operation of the softswitch involves a few steps or a few clicks to accomplish a specific task. The less time it takes, the better it is. Not just from the time-saving aspect, but also from the human emotions perspective, as users are accustomed to the comfort of operating a convenient and efficient piece software. Daily operations of the softswitch should be optimized to avoid manual work or, in ideal conditions, fully automated, so users can concentrate on their main tasks. During the evaluation, pay attention to your must-have features. Are they automated, semi-automated, or handled manually? The more time the softswitch saves you, the better business optimization you will achieve.

Stability

Because each softswitch has some capacity limits, it is quite important to make sure that the system you evaluate will perform well, according to the terms described by the supplier. You can get an idea about stability by:

- Using the system in a production environment. This option can reveal the most accurate results, but normally no one wants to experiment with their clients as it is too risky.
- Making a virtual stress test environment (there are special testing tools for SIP traffic generation, such as SIPp – a SIP performance tester). It is more secure compared to the first option, as you do not need to involve your

clients. However, a virtual environment usually does not encounter all the factors that appear in live traffic.

- Contacting bigger clients of your chosen supplier and asking them about their experience related to the system performance. It is the simplest approach and is recommended as a starting point. If you talk with at least five clients, you will get a clear picture about your chosen product and supplier.

Security

Security is a topic that requires more attention because of widespread VoIP hacking. Hackers can cause huge damage if the system is not well protected.

It is important to understand what security aspects are applied in the softswitch you evaluate. Here are some of the features that may be inbuilt in softswitches:

- Automated, unauthorized IP address-blocking to prevent hacker attacks using SIP scanning tools. (The softswitch automatically blocks an unauthorized IP address if someone tries to access the system a few times in a row.)
- Limiting the number of authorization or registration attempts to prevent the system from overload.
- Automated alerts to inform the administrator about possible threats or recommending them to take action automatically to prevent the system from damages.
- An online backup system that automatically saves the data and configurations on a remote cloud server when softswitch is the least loaded (e.g. every night at midnight).

- An inbuilt firewall or SBC function to ensure that your network is always available, secure, and fully manageable.

Here are a few additional recommendations that will help prevent hacking or at least reduce the possibility of becoming a victim:

- Choose strong passwords (many solutions are available to generate high-level passwords automatically).
- Deploy only trusted VoIP systems with the high security level. (Avoid connecting open source VoIP systems—Asterisk, FreeSwitch, FreePBX—to the public Internet, unless you have a team of experts).
- Make a virtual private network (VPN). VPN allows the sending and receiving of data between two or more devices across the public network as if they were directly connected to the private network, while benefiting from the functionality, security and management policies of the private network.
- Use VoIP encryption. You can enrypt calls by using specific products that support ZRTP or other encrypted protocols (e.g. Skype has inbuilt proprietary encryption).
- Use a session border controller or a SIP proxy as a kind of firewall for VoIP network. Add other elements that may increase security (firewall, anti-virus, or anti-spyware software).
- Block expensive destinations unless you specifically require them to be unblocked (like premium numbers and satellite phones).

- Allow a reasonable credit limit for your users; avoid setting unlimited call capabilities.
- Make sure your system is monitored 24/7. If you don't have such capabilities, use automatic alerts to notify you or your technical team about possible threats.

Support Level

Everyone dreams about solutions that do not require support at all. However, VoIP is technology-driven business, so technical support is especially important. If possible, evaluate the support level before investing in the softswitch. Usually support conditions will be described in a service level agreement, but in general the most important factors to consider are:

- **Response time.** It usually depends on the size of the technical team, the number of the active clients they serve and the frequency of reporting problems. The faster a response you get, the more confident and safe you feel.
- **Average time to resolve a problem.** Response time is more important for psychological reasons as it provides confidence that there is a person nearby who can take care of the problem. However, a more important factor is how fast your problem is actually solved. It is always better to get things done in one shot without engaging in long communication with an engineer.
- **Number of layers or levels.** Bigger companies have multiple support layers, where the first level listens to complaints, registers a problem, and passes it to the next level. If you are a tech-savvy person, you may be anxious about such situations, but this is how it works

in bigger organizations. However, smaller suppliers have one level—technical engineers who will take care of the overall process from registering your problem to fixing it.

- **Professionalism of engineers.** In the previous example, it is clear that it is easier for a technical person to work directly with another technical professional who may skip the "dumb" questions that are used to get through the process of problem registration. However, support should not be "too technical." Engineers working in this position should be client-oriented and polite.

- **Support availability.** 24/7 support is a must-have for a softswitch provider. It can be organized in shifts or by doing "on-call" support, where responsible engineers are monitoring high-priority problems after official working hours.

- **Culture and language.** Everyone prefers to have assistance in their native language, provided by people from their country. However, the world is changing with globalization, and softswitch vendors may have a central support team that is able to communicate in multiple languages and serve clients from different countries. Some companies outsource the technical support to third-world countries, which can also cause inconvenience for some clients.

In addition to the support level, check if there is training available prior to or upon purchase of the product. On-site customized training is the best option, but it costs a lot if the supplier is outside your country. You may consider online training.

Pricing

You may wonder how much it will cost to buy or rent the softswitch, but this is too hard to give a straightforward answer as there are hundreds of softswitch vendors and each applies a different pricing structure. However, there are some common factors that influence the final solution price:

- **License purchase vs. rent.** License purchase means that you pay one time and become an owner of the license. Theoretically, most licenses are perpetual, but practically, each vendor continuously releases new versions. To make your system up-to-date, you'll need to upgrade it. Moreover, you'll need to take care of the server that would meet the supplier's recommendations. Consider this as an additional investment you'll need to make. Some suppliers offer a license buyout (leasing) that reduces the initial investment for a client and allows split payments for a few months. An alternative to the license purchase is a rental license. You can choose a rental on your server or a cloud-based softswitch as a monthly subscription. Both of those options usually come as a package, including everything you need: softswitch, technical support, and software upgrades.
- **Upgrade frequency.** You should learn how often a supplier releases new versions and how much the upgrade costs. If you purchase a license that does not include upgrades, you can multiply the upgrade cost by its frequency, and you will get a yearly upgrade fee. As mentioned, if you choose a hosted or rented option, most likely you'll have upgrades and support included in the monthly price.

- **Technical support type.** Usually there are at least a few types of support (e.g. basic, standard and premium). The better the support you want, the more it costs. Some suppliers can even do configurations in the softswitch for you, so you can calculate whether it is better to manage the system yourself, hire a professional, or outsource all operations to your vendor's technical team. You may also ask about the cost of training and how it is organized.

- **Add-ons.** You should understand which functions come as the main feature set and which functions can be extended by enabling extra modules.

- **Capacity.** Each softswitch has a capacity limit that can be regulated by the supplier or depend on business conditions. A supplier may charge you based on the number of concurrent calls (also called ports or channels), traffic volume (usually measured in million minutes per month), or softswitch element (where each element, such as GUI, DB, SIP proxy, or media server has a specific price). Another option is an unlimited license that allows you to grow as much as you can. Be careful with this one as it may include other hidden fees. In general, it is important to understand how many calls you can handle with a single license and what costs are involved if you grow your business or need to add extra elements.

- **Implementation.** You've learned that a softswitch may come as a single server solution or as a cluster. If you consider redundancy, you'll need at least a two-server solution. Clarify what implementations for redundancy and scalability are available, how much each of them costs, and how much you will have to pay for adding an extra element to your implementation.

- **Custom pricing.** If you miss some function, you may request your supplier to implement it later. This is not a service provided by all softswitch vendors, but you may ask about their policy on this topic.

Chapter 5.
Finding Partners and Suppliers

Once you've made your decision on the business model, it's time to organize a solid business environment by carefully choosing partners and suppliers.

Some suppliers, instead of focusing on only one product, offer an all-in-one solution, including other cross-category products or services that are needed to run VoIP business, e.g., a softswitch, a multi-tenant PBX, a softphone and a mobile dialer. This is why you may find some softswitch vendors who also provide mobile dialers or offer A-Z termination services. However, there are pros and cons behind this. A company that offers "everything" loses focus on a particular product, and they cannot be as good as their competitors that concentrate only on a specific niche. On the other hand, the one-stop-shop makes it easy for an entrepreneur to enter VoIP business.

If your supplier doesn't provide everything you need, maybe there is something they could recommend. You can later use those recommendations in research for other partners.

This chapter will lead you through the key resources required to run a VoIP business, and will explain how to find the right partners and suppliers for those resources.

Voice Traffic

The voice termination (voice traffic) provider is the key partner in a VoIP business that defines the cost and potential profit. In the retail business, VoIP providers may buy termination services from one or a few traffic suppliers whereas in the wholesale market, the number of interconnection partners can reach a few hundred. These figures reveal that there is no one supplier in the world that would be the best, according to all the factors. In this section you will learn where to find voice traffic suppliers and what criteria to use when choosing whom to work with.

Where to Find

The main places to find a supplier are: the database of a local telecom register, online directories, social networks and forums, search engines, telecom events, and recommendations. In this subsection you will learn more about each of those sources.

Local Telecoms

If you plan to offer local calling services or call termination in your country, consider interconnecting with the tier-1 or tier-2 carriers. Most probably you are aware of them or, if not, you can easily identify them by using search engines. Another way to get a complete list of top providers in your country is by contacting the telecom regulation authority. Each regulatory organization has a database that contains information about companies that are licensed to offer different telecommunications services (radio, Internet, numbering, and telephony). Such information is either publicly available at the authority's website or access to it can be granted in some other way. Some of the local telecommunications service providers are incumbent operators that were former regional monopolies on landline services. Others are VoIP companies, small cable operators, or wireless service providers.

Search Engines

If your services will be based on international calls, you can locate suppliers all over the world by using a search engine. Even though searching looks like an easy process, many startups still waste your time by trying to combine adjectives like "cheapest" or "best" with keywords "VoIP provider" or "termination provider." As a result a search finds companies that have dedicated the biggest marketing budgets to being optimized in search engines; it does not necessarily mean that they are the best companies. Pay attention to reviews in social networks, forums, groups, and independent blogs; by reflecting public opinion they will provide far better recommendations.

Online Directories

A faster method to get a list of providers is to use an online VoIP directory. There are a few websites which have gathered needed information into one place and made it available for everyone. Below is a list of few online directories to visit:

- **VoipProvidersList.com.** This is one of the biggest VoIP directories and contains over 10,000 VoIP providers worldwide. They can be found under the "Providers" page and are listed by country. When you click on the description link, you'll find useful information, including address, responsible person, and their contact details.

- **Voip-Info.org.** A reference guide to all things about VoIP that also contains lists of many wholesale VoIP providers with short descriptions about them. They can be found in the following sections: "VOIP Service Providers," "VoIP Wholesale," "VoIP Termination," and "VoIP Routes." Each page can be edited by users, so the site does not look too user-friendly. But aside from that it is still a great resource.
- **Voip-Catalog.com.** A catalog of VoIP suppliers, containing information about more than 1300 wholesale VoIP providers.
- **Voip-List.com.** A catalog of VoIP suppliers, containing information about 500 international VoIP wholesale providers.
- **MyVoipProvider.com.** Includes the top 100 VoIP providers, ranked by user reviews, market segment, provider size, and price. However, most of them are retail VoIP providers. There's a separate section for wholesale VoIP which can be found in the "VoIP Wholesale" subsection on the "VoIP Provider Directory" page, but it contains just around 70 records.

Social Networks and Forums

The best social network for industry professionals worldwide is LinkedIn. It is a great place to find complete profiles of your contacts. If you want to review profiles of people outside your network, consider getting the Premium LinkedIn account which comes with some value-added features. However, the first recommendation is to join a few LinkedIn groups that are associated with VoIP.

Before joining a group, pay attention to the number of

members. This number will show how popular this group is (tip: to evaluate group activity, check the number of new conversations per day). Once you've joined the group, you'll find a complete member list on the "Members" page. As a group member, you'll be able to send a private message, even if the member is outside your network. (It's important to note that LinkedIn changes its privacy policy from time to time.) You can interact with other members by asking recommendations of wholesale VoIP providers publicly or in private by contacting more active group participants.

Additionally, there are many forums where providers can place their buy and sell ads for VoIP routes. I don't consider forums as a reliable source for finding a wholesale VoIP provider, but it's still worth mentioning as one of the alternatives. Here are few of the forums (start with those that have most members):

- Voipforums.com (over 60,400 members)
- Voipjungle.net/forum (over 25,000 members)
- Voipact.com (over 5,500 registered users)
- Calltermination.com (over 3,300 members)
- Voipbiznessforum.com (over 2,000 members)

Telecom Events

There are many telecom conferences, tradeshows, exhibitions, and workshops that are organized in different parts of the world. Some are dedicated specifically for carriers who want to exchange traffic in wholesale. However, it is the most expensive option; you are required to buy a flight, book a hotel, and invest one or two days of your time. On the other hand, during such events you can meet representatives from the main telecom operators directly. After meeting a person, it's much easier to develop a business relationship further.

If you want to participate in telecom events, prepare for them in advance and do your homework. Make a list of suppliers who you want to talk to and schedule meetings with them. In your spare time, you can visit other suppliers as well, but try to focus on those who are capable of becoming your partner. If you are just a beginner, don't expect something from just one single carrier; it is better to concentrate on those suppliers who aggregate traffic from smaller companies like yours and then pass it to bigger telecoms. Here are the most popular events, where you can meet voice traffic suppliers and other companies, working in telecom industry:

- **AfricaCom.** The biggest tech event in Africa that gathers together senior decision-makers from the entire digital ecosystem with 10,000+ attendees, 350+ speakers, and 375+ exhibitors. It is organized in Cape Town, South Africa.
- **Asian Carrier Conference.** The telecom industry event, attended by IT and telecommunications professionals from Asia and the rest of the world.
- **Capacity Conferences.** Capacity Media organizes around 20 conferences worldwide each year (including ITW, which will be mentioned later in this list). Conferences are focused on the wholesale telecommunications market and carrier-to-carrier business.
- **CommunicAsia.** An information and communications technology (ICT) exhibition and conference, held in Singapore.

- **Comptel Plus.** The networking event in the USA for telecommunications industry professionals looking to expand their networks and strengthen existing relationships.
- **Gitex Technology Week.** The most widely recognized and annually anticipated ICT trade event across the Middle East, Africa, and South Asia. Gitex is held in Dubai, the UAE.
- **Global Carrier Community Meeting (GCCM).** There are a total of six events worldwide, organized by Carrier Community. You can meet the Club Members representing decision-makers from the tier-1, tier-2 and tier-3 operators worldwide.
- **International Telecoms Week (ITW).** The world's largest meeting for the global wholesale telecommunications community, held in the USA. ITW provides the platform to schedule meetings and offers a unique blend of networking and business opportunities.
- **Internet Telephony Expo (IT Expo).** The event in the USA with an educational program that teaches resellers, enterprises, SMBs, and government agencies how to select IP-based voice, video, fax, and unified communications to purchase or resell. ITEXPO is where buyers, sellers, resellers, and manufacturers meet to forge relationships and close deals.
- **Mobile World Congress (MWC).** The world's largest gathering for the mobile industry, organized by the GSMA, held in Barcelona, Spain.
- **TelecoDays.** A networking event uniting wholesale telecommunications experts from around the globe. Currently it is held three times per year in different cities: Prague, Las Vegas, and Dubai.

- **Telecom Exchange (TEX).** Brings together leading buyers and sellers of wholesale data and transit providers, who need to connect to decision-makers, to negotiate and close business deals. It is held in London and New York.
- **Wholesale World Congress (WWC).** The meeting point in Madrid, Spain for the international telecoms wholesale community including tier-1, 2 and 3 carriers, mobile operators, ISPs, VoIP providers and technology partners from the voice, data, satellite and fixed-line markets. Organizers of WWC also have two more events: America Wholesale Congress (AWC) in Miami, the USA and Latin America Wholesale Congress (LAWC) in Buenos Aires, Argentina.

Online Minute Exchange

It was mentioned earlier that VoIP traffic can be treated as a commodity that can be evaluated by specific quality benchmarks. If there are similarities, can it be traded in an online marketplace publicly? Does such a platform exist in general?

There have been many projects to accomplish this, but most of them have failed. Whenever other wholesalers are asked if they have tried to use such minute trading systems, the answers are either negative or that they have tried, but the volume was relatively small—just one or two percent of their overall traffic. They explained that such marketplaces were not competitive. Good examples are not available, but you may be able to find some companies offering an online marketplace. However, you may also find out that some of them are simply wholesale carriers looking to sell and buy routes in this way.

Recommendations

Carrier-to-carrier business is all about trust, credibility, and relationship management. Your friends and partners will recommend those providers they work with and trust, so recommendations are probably your most important source of all. If you do not have friends or partners who work in the VoIP market, you can use an alternative method—ask for references in social networks, LinkedIn groups, and online VoIP forums.

With many available options for finding providers, the question remains over how you choose which provider is right for you. The next subsection helps you to understand the criteria that differentiate suppliers from each other.

NOTE FROM THE AUTHOR: I've noticed that most starting VoIP companies choose VoiceTrading as the traffic supplier. (The original name of this company was Dellmont Sarl.) This company manages many "clones," such as Siptraffic, VoipPro, and Betamax that provide either retail or wholesale VoIP services. Even though each of those brands belongs to the same company, calling rates are different. If you search for "Betamax clones," you'll find a lot of information about this company including the portal that lists all brands and gives a call rate comparison table.

The main reason that VoiceTrading has been chosen by many entrepreneurs is very simple—low rates. However, do not be distracted by this. Read the next subsection carefully and compare them in the same way as you should with other providers.

How to Choose

Price and quality are not the only benchmarks that can be used to compare VoIP traffic providers. In addition, there are plenty of important aspects to take into consideration.

Experience

It is always better to deal with companies that have been in the telecom business a long time and have a strong brand. Such companies may not have the best rates, but you will have a credible and trusted partner. It does not mean you cannot work with smaller providers, but be aware of the risks. It's especially important because of widespread VoIP fraud. Fraudsters usually form new companies and may attract you with low prices, only to disappear after receiving a large prepayment from you. If you decide to deal with a company that is not well known, ask for references, check the LinkedIn profiles of its owners or employees, and negotiate payment conditions that will minimize your risk.

Provider Level (Tier)

Earlier it was mentioned that there are a few companies at the same level as the telecom operators. If you are a small fish, don't expect the whale to work with you. Big telecom companies usually have high voice traffic volume requirements and complicated interconnection procedures. It may take months to negotiate and sign an agreement with them. However, that does not mean there are no flexible tier-1 operators. Try to review their conditions and, if possible, have a chat with the responsible person before starting to explore other alternatives. If you do not meet their requirements or dealing with them takes too much time and energy, consider lower-tier operators or wholesale transit providers. They have interconnections with the largest carriers in different countries and can deliver affordable, fast, and flexible traffic aggregate services for smaller providers and resellers.

A-Z Rates and Key Destinations

A majority of transit providers offer A-Z rates that refer to termination to all worldwide destinations. Calls can be terminated directly or via a third party partner. A-Z termination is very convenient for retail VoIP providers because it is enough to have one supplier to terminate all worldwide calls. However, there is much competition in the global A-Z termination and transit operators have to find a niche to differentiate themselves from each other.

This is why wholesalers focus on termination to a few countries or a region by establishing direct connections with the first and second-tier telecom operators. Such interconnections guarantee the highest level of quality. But if the quality is not a wholesaler's priority, they may concentrate on the best route price by managing grey termination points. This allows transit operators to generate their best offerings, referred to as the "top routes" or "key destinations."

Whenever you approach a provider, ask about their key routes. If you hear a wholesaler saying they have "good rates" everywhere, most likely they are seeking to win a client by using any means possible.

Traffic Volume

Wholesalers may have standard pricing available on their website, but in general, rates depend a lot on call volume. Before sending you an offer, a provider will ask about your monthly traffic consumption and to which destinations you terminate those calls. Again, if you are just starting out, you will probably receive a standard pricing, but that is normal. The higher traffic volume you send through your provider, the better rates you can negotiate.

Traffic volume is important, and not only from the price perspective. Each supplier has some technical call capacity limitation that is determined by their hardware and the type of the traffic they accept. This is another reason they ask their clients about the traffic load and call volume. Make sure your supplier's limitation will not become your bottleneck. You need to know what call capacity your provider can accept. By knowing this you can configure the routing algorithms accordingly.

Price and Quality

Better quality comes at a higher price. You can request information about ASR, ACD, and CLI translation to compare it with others according to the price and quality ratio. In the table below you can see a sample rate sheet of key destinations:

Destination:	Quality	ASR	ACD	Rate
Azerbaijan Mo-bile	NCLI	25-35%	1.8	$0.22/min
Ethiopia Mobile	NCLI	23%	2.5	$0.102/min
Ethiopia Mobile	CLI	56%	5.34	$0.179/min

Guinea All	NCLI	21%	1.56	$0.265/min
Italy Mobile WIND	NCLI	31%	3	$0.0119/min
Japan Mobile	NCLI	32%	2.13	$0.0236/min
Kazakhstan Mobile Kartel/Kcell	TDM	27%	2.65	$0.073/min
Kenya Mobile Safaricom	NCLI	20%	3.5	$0.05/min

The first column shows the country and the network type. You can see that rates can be the same for all networks (Guinea All) or specific for the certain operator (Mobile WIND, Mobile Safaricom, or Mobile Kartel/Kcell). The second, third, and fourth columns show the quality of calls; the last one—price.

Once you agree on price and quality terms, the next step is to make some real tests by sending real traffic to this carrier. Tests will show whether the quality parameters of the offering are correct.

Route Testing

Most providers will give you free credit to evaluate the quality of their routes. If not, you may be able to prepay a minimal amount to make some test calls. You can perform testing by adding a provider to your softswitch and making calls to different destinations by yourself.

An even better method is to open a provider for your clients to evaluate and collect their feedback. In this case, it is better to test with close and friendly customers who will not be so sensitive if case quality is below a satisfactory level. After placing some calls, review reports of the main quality benchmarks and listen to a few conversations. Listening to live calls or recordings can be done randomly or by checking only suspicious calls that have low ASR or ACD.

It is also important to note that sometimes providers give a better quality termination during the testing stage to attract a potential client. After that, they start sending your traffic to the low quality routes. To avoid this, you must enable an automated quality monitoring that would quickly notice any deviation from the standard. You'll easily recognize which providers offer reliable and stable voice traffic in the long run. Such suppliers usually have network redundancy, traffic with minimal downtime and many alternative routes if the main one fails.

Payment Conditions and Methods

Most providers deal both in prepaid and postpaid, but each of them has their own payment terms and conditions. Normally, everyone starts relationships with the prepayment option, especially when two companies negotiate for the first time. After working together for a while, they may switch to postpaid.

First, ask your provider about the minimal voice traffic purchase amount. Looking at the the long-term perspective, at some point a provider may give you credit for a week or month. Try to understand a provider's policy by asking the right questions about payment conditions before establishing a partnership with them. For some suppliers it may be enough to provide strong references; others base their decisions on credit ranking agencies or on personal experience, which requires an evaluation of a new partner's credibility for a specified time.

A provider may have different methods for accepting money, such as a wire transfer, Paypal, Western Union, or Skrill (formerly Moneybookers). Each of the payment options has a different commission rate that can be based on a percentage, fixed fee, or a combination of both. You need to add this to your financial calculations and find the most convenient method to transfer funds to your supplier.

Also, it is worth mentioning that sometimes route buyers and sellers are not willing to accept the risk of working together. A buyer may be afraid that proper service will not be delivered after making a big prepayment. A seller may be cautious about the buyer's ability to pay on time. In this case, both parties can consider dealing through a clearing-house or an escrow service company that will manage neutral minute exchange transactions. Such partners protect both parties and reduce their credit risk in exchange for commissions from each transaction.

Bilateral Interconnection

If you have some good routes to offer your supplier, you may consider a bilateral interconnection. It is a very attractive option, since you do not need to transfer funds to each other very often, and you will save on transaction fees. It works like a barter, e.g. you give 120,000 minutes to Argentina in exchange to 10,000 minutes to Armenia. In this example, two particular destinations are used, but normally both parties can call to any destinations. Each week or month, providers check the bilateral balance and issue an appropriate invoice to each other.

Support

The last thing to mention is customer support. However, it

was reviewed in detail in the previous chapter (see the "Evaluation" subsection of the "Comparison Elements" section). When you evaluate support, pay attention to the average response time (it is more relevant for client control and confidence), the actual time of resolving the problem, and the availability of technical personnel (it should be 24/7).

NOTE FROM THE AUTHOR: If you took every single provider in your list and went through each of the points listed above, it would take forever. What to do instead? I'd personally recommend making a decision based on two main factors— references and personal experience. In addition, always be ready for changes and new opportunities. Actively work with your suppliers and ask for volume discounts if your traffic increases. Finally, it's natural that if you expand your business, at some point you may outgrow your supplier. If you think that you are not getting a proper price, you can always evaluate other options in the market.

DID Numbers

There are three main choices for obtaining DID numbers. First, you can get a specific license, and a number range will be assigned by the regulation authority. Second, you may acquire DIDs from a provider which already has such a license and a number range. Third, you can buy DIDs from the DID aggregator.

Each country has an institution that regulates telecommunications services. It is responsible for allocating a range of specific numbers for a licensed operator. From here, an operator can supply allocated numbers to end-users as a part of their service or resell DIDs to other smaller providers that do not have a license or a number range. Usually it does not cost anything to obtain a range of numbers for licensed operators, except for licensing fees which are relatively small in developed countries. If you need only local DID numbers, research the process of obtaining a number range from local authorities. In this way you will be able to evaluate whether it is worthwhile to obtain the necessary license and invest your time. However, the self-cost of DIDs is relatively small, so a majority of VoIP providers choose to purchase DIDs from local providers or DID aggregators. DID aggregators (also called DID wholesalers) supply DID numbers in bulk, worldwide. It is the most convenient service for those who require DIDs in a few countries.

You can find DID aggregators in search engines by using the following terms: "International telephone numbers," "VoIP origination," "DID numbers," "Virtual numbers," "Wholesale DID," and "International DIDs." What follows are examples of worldwide DID aggregators.

DIDX
Has numbers from more than 65 countries and 9,000 cities with instant activation and low monthly rates. These numbers can be used for calling cards, callback, call forwarding, PBXs, mobile phones, voicemail, faxing, or other voice services. Numbers are available on a basis of per-channel, per-minute or per-trigger.

DID LOGIC
A DID trading platform with SIP, PSTN, Skype, and Gtalk forwarding features. It has a massive stock of the UK DID

numbers from seven nationwide carriers. The total number coverage is from more than 50 countries. Numbers can support up to 500 channels and prices start at $0.12 per month.

DIDWW

An inbound DID carrier providing origination services from more than 60 countries worldwide. A company which works with local and global telecom service providers, mobile network operators, integrators, and voice application providers.

Voxbone

Specializes in inbound communications and delivers geographical and toll-free DID numbers from more than 50 countries around the world. The company enables cloud communications providers, global carriers, and enterprises to extend the reach of their voice services internationally. Voxbone services are real-time and fully automated.

When selecting a DID supplier, pay attention to the number of channels available in a DID and whether this number can be expanded or reduced. It is hard to predict how many channels you'll need at the beginning of your services, especially if you deliver calling cards. You would not like the number of DID channels to be your business bottleneck, so it is recommended that you consult your DID provider and ask if they've had a case identical to yours. If you choose to work with a bigger DID aggregator, it is likely they have had a provider similar to you and will be able to offer recommendations. It is also important to note that if you want to use a callback, make sure your DID provider sends the correct caller ID information. Without a proper caller ID translation, your system will not be able to determine who to call back.

Software and Hardware

Most of us choose to buy products or brands that we know, like, and trust. However, vendors with established brands and higher positions in the market are usually more expensive. This encourages some of us to look for more affordable solutions.

During my work with VoIP providers, I have noticed that most of them choose the vendor based on recommendations from friends or business partners—people they know and trust. However, if you do not yet work with people connected to the VoIP industry, you may ask for referrals in LinkedIn groups or by contacting some companies directly. Once you have a list of a few vendors on it, request references and testimonials from their clients and call them to confirm the credibility of the supplier you evaluate. You may also find much useful information by using search engines. Most of the suppliers are visible on the Internet, and you may find responses about them in various social networks, forums, and groups.

This section reviews vendors that provide software and hardware for VoIP business.

Open Source Platforms

Open source software is chosen because of two main reasons: to save money and to be able to customize it. This subsection reviews different types of open source platforms, including SIP servers, softswitches, and PBXs.

Asterisk

Probably the most popular framework for building a communications application that can turn an ordinary computer into a PBX. Asterisk was created in 1999 by Mark Spencer and is sponsored by Digium. Like any PBX, it allows attached telephones to make calls to one another and to connect to other telephone services, such as the PSTN and VoIP. Asterisk is released under a dual license model, using the General Public License (GPL) as a free software license, and a proprietary software license to permit licensees to distribute proprietary, unpublished system components.

Asterisk is the heavyweight of open source PBX platforms and it is a core component of many PBX-in-a-box commercial products and open source projects (A2billing, FreePBX, and Elastix). Some of the commercial products are hardware and software bundles, for which the manufacturer supports and releases the software as open source.

A2billing (Asterisk2Billing)

Probably the most popular open source Class 4/5 softswitch with inline billing. A2Billing is used for customer provisioning and management. You can use it as an all-in-one solution for providing many different telephony services: residential, SIP trunking, wholesale transit, DID resale, calling cards, and callback.

Commercial installation and support of A2billing are available from Star2Billing S.L. It also provides training, consultancy, and additional value-added products and services for the telecom industry.

FreePBX

An open source web interface (GUI) which controls and manages Asterisk. FreePBX is a very large part of why Asterisk has been so successful. The FreePBX interface allows you to configure and manage many aspects of an Asterisk system without touching a single configuration file. The FreePBX distribution includes packages that offer VoIP, PBX, fax, voicemail and email functions. It is easy to customize and adapt to changing needs. FreePBX can run in the cloud or on-site and is currently being used to manage the business communications of all sizes and types of businesses from small one-person SOHO businesses, to multi-location corporations and call centers.

FreePBX is licensed under GPL and is a registered trademark of Sangoma Technologies. The main sponsor of FreePBX and the FreePBX.org project is Schmooze Com Inc., a Sangoma company.

Elastix

Open source software for unified communications; it incorporates all the communication alternatives (PBX, email, instant messaging, faxing, and collaboration function) into a unique solution. The Elastix project began as a call report interface for Asterisk and was released in March 2006. Later that year the project evolved into an Asterisk-based distribution. Currently Elastix is maintained by PaloSanto Solutions based in Ecuador and uses a forked version of a FreePBX user interface.

Users sometimes confuse distributions of unified communications with telephone exchange systems (PBXs). Elastix not only provides telephony, it integrates other communication alternatives to an organization environment more productive and efficient.

trixbox

PBX software, based on Asterisk. trixbox CE was available under a GPLv2 license as the community edition and was completely free. However, in 2012 Fonality (the current owner of the trixbox brand) announced that they would no longer support or update trixbox CE, and would instead focus on trixbox Pro—the commercial version of trixbox. Unlike the community edition it contains components for which the source code is not available.

FreeSwitch

FreeSwitch is a scalable open source telephony platform designed to route and interconnect different communication protocols using audio, video, text, or any other form of media. The project was initiated to focus on several design goals, including modularity, cross-platform support, scalability, and stability. FreeSwitch can be used as a softswitch, proxy server or an SBC. It provides a stable telephony platform on which many applications can be developed using a wide range of free tools. (It is also worth mentioning that all three FreeSwitch creators were former developers of Asterisk.)

Asterisk, A2billing, FreePBX, Elastix, trixbox and Freeswitch are probably the most popular open source platforms, but actually there are many more. Keep in mind that it is not easy to maintain a free community edition and as a result, some of the projects are discontinued or stopped. The more successful projects are often purchased by bigger organizations and turned into proprietary systems (e.g. as it has happened with trixbox) or commercialized in some other way. Here is a list of few additional open source softwares that you can evaluate:

- **AsteriskNOW.** A "packaged" solution that includes Asterisk, the DAHDI (Digium/Asterisk Hardware Device Interface) driver, and FreePBX. DAHDI is the open source device interface used to control Digium and other telephony interface cards (for interoperability between VoIP and legacy telephony).
- **AskoziaPBX.** An open source PBX based on Asterisk. However, since the 2.1 version, it has been released under proprietary license only.
- **FusionPBX.** An open source web interface (GUI) for FreeSwitch. FusionPBX can be used as a standalone or a multi-tenant PBX, softswitch, call center solution, SIP server, or voice application server (for voicemail, fax, or conference calling).
- **SEMS** (SIP Express Media Server). Free software, licensed under dual license terms (public and proprietary license). It is intended to complement proxy servers and works as a media and application server for SIP-based VoIP services.
- **Kamailio** (formerly OpenSER). An open source SIP server released under GPL. It is able to handle thousands of call setups per second. Kamailio can be used to build large VoIP servicing platforms or to scale up SIP-to-PSTN gateways, PBX systems, or media servers such as Asterisk, FreeSwitch, or SEMS. Kamailio and the SIP Express Router (SER) teamed up for the integration of the two applications and new development.

- **OpenSIPS.** A branch of the SER project, offering an open source SIP server for presence, voice, video, and IM. OpenSIPS is intended for installations serving thousands of calls. It is a multi-purpose solution that can act as a SIP router, application server, load balancer, NAT traversal server, or IP gateway.

- **SipXecs.** An open source VoIP server. In terms of the features, it is not unlike Asterisk, but the design of SipXecs deviates from Asterisk in many ways. Development started in 1999; in 2004 Pingtel Corp contributed the code base to the non-profit organization SIPfoundry. SipXecs includes many features of a traditional PBX, such as voicemail, IVR and auto attendant. The main components of the system are designed around FreeSwitch.

- **Vicidial.** An open source platform for call centers. Vicidial is designed to interact with the Asterisk to act as a complete inbound and outbound contact center suite. The project is maintained by the Vicidial Group.

- **Yate** (Yet Another Telephony Engine). Free and open source communications software with support for video, voice, and instant messaging. It was created with a modular design in mind, allowing the use of scripting languages to create external functions.

Proprietary Platforms

Open source platforms are not suitable for everyone, especially those who are not tech-savvy or do not have a skilled technical team. Sometimes it is better to invest more and get a solution which works properly and comes with initial training, professional technical support, and regular updates. This subsection lists vendors who provide proprietary softswitches, PBXs and unified communication platforms.

Below is a list of vendors providing proprietary softswitches. (It is important to note that softswitch may be just a part of their product portfolio, so do visit their websites or contact representatives for more details.)

- **Aloe Systems (SwitchRay).** A developer of VoIP softswitch solutions, session border controllers, and service platforms for telecom carriers. A company well known for its transit switches: MVTS Pro, MVTS I, and MVTS II that are still widely used for wholesale business. In 2013, all intellectual property rights (including products) of Aloe Systems were acquired by SwitchRay.
- **Digitalk.** A vendor of carrier, consumer and mobile communications solutions to retail and wholesale service providers. The company offers complete and scalable platforms with applications and integrated management to operate different types of telecommunications services.
- **Integrics.** A software engineering company providing telecommunications solutions. Integrics is the developer of Enswitch, a softswitch and hosted PBX platform which allows managing a wide range of commercial telephony services such as: multi-tenant PBX, toll-free and number translation services, calling cards, and call shops.
- **IPsmarx.** A supplier of Class 5 softswitch, billing and prepaid solutions for VoIP service providers. Covers many different services, such as international mobile

top-up, pinless dialing, calling cards, and mobile VoIP.

- **Kolmisoft.** A developer and provider of Class 4/5 softswitches with inbuilt billing. The company is best known for its all-in-one solution, MOR, which has modular architecture and can be used to deliver the following services: residential VoIP, SIP trunking, hosted PBX, calling cards, callback, call termination, and wholesale transit. Another product of Kolmisoft is M2, a Class 4 softswitch for wholesale transit (IP-to-IP interconnections). (**Disclosure:** *I work at Kolmisoft.*)

- **Linknat (Kunshi Network Technology).** A provider of various software solutions to manage a VoIP business. They are best known for their wholesale transit switch, VOS3000. It is widely used by service providers in the Middle East and South East Asia.

- **PortaOne.** A software developer of billing and Softswitch solutions for telecommunications service providers, operators, and carriers. VoIP providers can run a broad line of retail and wholesale services within a single software package. PortaOne provides only on-premises software; for cloud-based solutions it partners with other companies. This is why you may find different suppliers that offer PortaOne switch partitioning services as a white-label solution.

- **REVE Systems.** A software solution provider with a focused approach to serve the IP-based communications industry. The company specializes in Class 4/5 softswitches (iTel Switch and iTel Billing), mobile dialers (iTel Dialer) and bandwidth optimization solutions (iTel Byte Saver). It mainly serves clients in the Middle East and South East Asia.

- **Sippy Software.** A software development company that

delivers VoIP solutions to provide wholesale and retail telecommunications business. They provide Class 4/5 softswitches to manage different types of business models: wholesale, call center, DID arbitrage, calling cards, pinless dialing, callback, switch partitioning (white-labeling), and call shop.

- **Sipwise.** A software developer and integrator of carrier-grade Class 4/5 softswitches for all kinds of access networks. Has a free community edition as well as commercial products for call management, switching, and unified communications in the cloud.
- **Speedflow.** A telecom and software solutions provider. The company provides Class 4/5 softswitches, VoIP ERP system, transaction processing solutions and voice traffic.
- **Voipswitch.** A VoIP software developer (a subsidiary of Voiceserve) that provides a Class 5 softswitch (Voipswitch) with integrated billing, user portal, mobile and desktop dialers, and rich communication suite.

Additional vendors that were not mentioned here, but deliver softswitches, SBCs, and media gateways: Dialogic, Cisco, Genband, Nexge, Quintum, Sansay, Squire Technologies, TelcoBridges, Teles, and Topex.

It is important to note that not all softswitches come with an inbuilt billing system. As an example, there are many transit providers that are using MVTS II as a switch and JeraSoft (developer of telecom billing software) as a billing system. When contacting softswitch vendors, ask if the billing is inbuilt. If not, what billing solution do they recommend?

Next is a list of few PBX solution vendors:

- **3CX.** A provider of Windows-based IP PBX that completely replaces a traditional proprietary phone system.
- **4PSA.** A provider of a cloud PBX and unified communications platform, named VoipNow, which is dedicated to service providers and organizations. Providers use it to deliver cloud services to SMBs and enterprises. Organizations use it as a PBX to manage voice, video, presence, instant messaging, conferencing, faxing, and mobility.
- **Bicom Systems.** A solution provider of proprietary PBX (PBXware) for SMBs, enterprises, governments, and service providers. The company also delivers other solutions and applications to manage communications.
- **Brekeke.** A developer of SIP software products for IP network communications. They provide SIP proxy server and IP PBX software for a wide variety of service providers and enterprises.
- **BroadSoft.** A provider of software and services that enable mobile, fixed-line and cable service providers to offer unified communications over IP networks. The company's core communications platform enables the delivery of a range of enterprise and consumer calling, messaging and collaboration communication services, including PBX, video calling, text messaging and converged mobile, and fixed-line services.
- **CommuniGate.** Develops carrier-grade unified communications software for broadband and mobile network operators to deliver cloud-based services to business subscribers.
- **Dialexia.** A telecommunications software company

specializing in the development of PBX software and VoIP business solutions for corporations, small businesses, service providers, and telephony resellers. The company also provides custom-built PBX applications, VoIP softswitch, telecom billing, and delivers professional services.

- **Digium.** The company behind Asterisk, which provides a wide range of products. One of them is Switchvox, an IP PBX that delivers powerful unified communications tools, mobility applications, and robust calling features.
- **Thirdlane.** A company provides multi-tenant IP PBX and unified communications solutions for businesses, VoIP service providers, and call center operators.

Some of the softswitch and PBX vendors provide solutions that are based on open source platforms (Asterisk, FreeSwitch, or OpenSIPs); others have built their own customized telephony engines. You may notice that proprietary solutions, based on open source technology are usually more cost-effective. They are preferred by those people already familiar with open source software, who are looking for something that is more robust and stable.

In addition to open source and commercial solutions, there is also a "black" market of cracked proprietary softswitches. You can find lots of people selling cracked softswitches by announcing their posts in various VoIP forums. Even though you may be attracted by their prices, this is not recommended; you could damage your reputation and risk your business on a cracked solution.

Devices

There are many manufacturers who produce telecommunications hardware and software. Most probably you've heard of brands such as Avaya, Cisco, Sipura/Linksys, Nortel, Polycom, Alcatel-Lucent, Grandstream, and Yealink. If you visit their official websites, typically you will find a list of their resellers and distributors. Companies which deliver telecommunications products in specific regions usually represent several brands.

Once you've found a local equipment distributor, request that someone assists you in making a decision about VoIP equipment. Then explain your business case and ask for an opinion. Most representatives will be glad to consult with you about different types of products and will guide you through the process of selecting the best one for your scenario. You may organize the same dialog with two or three distributors to gather the most relevant information and form a strong opinion on your own.

In addition to this, compare equipment prices in an online marketplace, such as Amazon, eBay, or Alibaba. If you notice that other sellers offer lower prices, use this as a method to negotiate with the distributor.

Normally it is better to deal with distributors; they usually provide professional services, which is especially useful if you don't have much experience in telecommunications. Some suppliers position themselves as industry experts and provide useful resources, video lessons, and white papers about choosing the right VoIP software and hardware.

A good example is a company named VoIP Supply which issues various editions of VoIP buyers' guides. They include the main aspects to consider before investing to VoIP equipment. Once you've chosen products that you'd like to use, ask the distributor if it is possible to test the equipment in your network. If this is not possible, it is good practice to search for feedback from other VoIP equipment users. You can easily find such information in various forums (Webhostingtalk.com, Dslreports.com), social networks, groups, and blogs. If you do not find the right answer, you may open a new topic by yourself and ask to share the experience with specific vendors.

Applications

You've already learned that there are desktop softphones used by those who prefer calling from a fixed location (home or office) and mobile dialers that run on smartphones and allow calling from anywhere using mobile data or a Wi-Fi connection. Those applications are available to download directly from a supplier's website or an application distribution platform (iOS App Store, Google Play, Amazon Appstore, Windows Store, or GetJar). This subsection explains how you can cooperate with an application developer to be listed as a VoIP provider on their softphone and how to buy a ready-to-use app.

The first option is to cooperate with an application developer. You need to visit a few popular application distribution platforms and research the top telecommunications apps. Review the apps under "communication" or a similar category or search by using specific keywords like "voip," "softphone," "sip phone," or "mobile dialer." In addition to this, you can easily find a list of app names by using search engines. There are a number of websites and blogs that compare the most popular or the most downloaded VoIP apps.

When you open a description page of an app, you'll find information about the number of downloads, user ranking score, name of the supplier, and other useful information. Most probably you'll recognize the most popular brands like Facebook Messenger, Viber, WhatsApp, Skype, Google Hangouts, Nimbuzz, Tango, Rebtel, Voxox, RingCentral, and Vonage. In addition, you'll find many names that do not sound familiar. Try to download and evaluate those apps by yourself.

Some of them belong to service providers that use them as their main product to deliver communication services (you can treat them as your competitors if you plan to operate in the global market). Others are created by software developers and allow a choice of multiple pre-set VoIP providers, or offer the facility to add details of a SIP account manually. Those are the apps that you should be looking for.

Once you have found a few popular apps, contact the developers and ask about the terms, conditions, and procedures for being listed as one of the pre-set providers in their application.

Below is a list of some popular open source and proprietary VoIP softphones. Not all of them have cross-platform capabilities; before downloading them, please visit the supplier's website and check the compatibility with your operating system.

- **Blink.** A real-time communications client for audio, video chat, presence, file transfer, and screen sharing using SIP protocol. Available for Mac, Windows, and Linux. Blink Pro version is available in the Mac App Store.

- **CSipSimple.** An open source VoIP application that runs on Google Android operating system and uses SIP for calls and instant messages. It allows conference calls, secure calls, transfers, and has an echo cancellation feature.
- **Ekiga (formerly GnomeMeeting).** An open source softphone, video conferencing and instant messenger application. It supports HD sound quality and video up to DVD size and quality. Ekiga uses the major telephony standards (SIP and H.323); it is interoperable with most of the standard compliant VoIP software and hardware.
- **Jitsi (formerly SIP Communicator).** A free and open source voice, video conferencing and instant messaging application for Windows, Linux, Mac OS X, and Android. It supports the SIP protocol for calls and instant messaging and allows secure encryption for voice, video, and chat.
- **Linphone (contraction of Linux phone).** A free SIP client initially developed for Linux but now supporting many additional platforms, including Microsoft Windows, Mac OS X, and mobile phones running Windows Phone or iOS. It supports SRTP for end-to-end encrypted voice and video communication.
- **MicroSIP.** A free and open source softphone for Microsoft Windows. It facilitates high quality VoIP calls based on the open SIP protocol.
- **SFLphone.** A free and open source SIP and IAX2 com-

patible enterprise softphone for desktop and embedded systems. It is released under the GPL; packages are available for all major Linux distributions.

- **Sipdroid.** A SIP application for the Android operating system. Sipdroid interfaces with Android's default dialer application and optionally prompts the user to make an outgoing call using Sipdroid or the GSM/3G network. It supports two simultaneous SIP accounts, STUN (session traversal utilities for NAT) for users behind NAT and video calls.
- **Twinkle.** A SIP softphone for VoIP calls and instant messaging. It can be used for direct IP-to-IP communication or in a network using a SIP proxy to route calls and messages.
- **X-Lite.** A proprietary freeware SIP-based softphone, developed by CounterPath Corporation. They also provide other desktop and smartphone clients—Bria and eyeBeam.
- **Zoiper.** A softphone to make VoIP calls through the IP PBX or SIP provider. Available for iPhone, Android, WP8, Windows, Mac and Linux.

Quite often, as well as the free application, a supplier provides a "Premium" or "Pro" version that is more feature-rich and comes as a paid option. Please also note that the list above is not the official list of suppliers to contact. This list will help you to get started, but you will have to do additional research.

Even though being listed as one of the pre-set providers gives you a great chance of market exposure (especially if you are just starting), it's not always the best way. Firstly, users won't see your brand as the main one on the app. It does not seem very professional because all major providers have apps with their own logo and company name. Secondly, it does not give you complete control of the business since the application does not belong to you and you cannot influence the value-added features in it. Finally, users may not choose you as their main provider for international calls, and you may lose business to other suppliers that are listed in the app.

To avoid this, consider another option: choosing a vendor that specializes in the development of softphones or mobile apps and buy a complete solution from them. Then you can customize it according to your brand and needs. Users will feel more confident about your brand when they see your logo, company name, and splash screen.

Freelance Development

There is a chance you will not find the right company for an application or for software which fulfills your needs and budget. Also, you may want to have something unique that is not yet available in the market. In both cases, you can consider hiring freelance developers to program new solutions, make additional customizations, or integrate different pieces of software.

You can locate talents in specific freelancer hiring portals such as Upwork (formerly oDesk), Elance (now an Upwork company), outsource.com, guru.com, peopleperhour.com, and freelancer.com. You may also try Fiverr, the marketplace for creative and professional services where basic gigs start at $5.

Hiring freelancers is more affordable and often faster

than buying services from the development company. There is a reason for this. You may find many blog posts of people complaining about their experience with freelancers. The fact is that it is difficult to control the remote work of freelancers and set common expectations, especially if this is your first project. Be ready to spend a lot of time and energy to organize this work. Read books that explain the proper way to hire and control the work of a freelancer (*oDesk Essentials, The Complete oDesk Handbook*).

When you start a new project, create the minimum viable product (MVP) first. It refers to the product that has the highest return on investment versus risk. (Read more on this in the book *The Lean Startup* by Eric Ries). Before creating something, talk to your potential clients (focus on decision-makers), and ask about their biggest pain. Then you can propose a solution by describing how your service would solve that. A simple conversation with your customers saves a lot of time, money, and resources of energy for you. Once you have a clear picture, you can create a sample prototype for your beta clients.

NOTE FROM THE AUTHOR: *I'm personally using a mobile app that is a perfect example of an MVP. I need a VoIP provider only for international calls. If I open an app (normally there's no reason to do that), there are just two possibilities— turn it on or off. In my case, it's always on. Whenever I dial an international number, the app recognizes the prefix and calls go through the DID number automatically. If calls are local, they are connected directly. For me, it's an ideal solution. I don't need to open an app, and I don't need an Internet connection; it does its main job without any extra features. I guess* *it did not cost much for my VoIP provider to develop it. So try to start with something simple that would contain only must-have functions.*

Finances

A VoIP business can be started by using your own capital or borrowing money from your friends or family members. There's even a well-known abbreviation FFF (friends, family, and fools) which describes the first-round financing sources for new startup companies. To achieve faster growth and more market exposure, consider finding a financial partner such as a private investor, business angel, venture capital, bank, or other financial institution. However, before approaching someone, you need to create a financial plan by calculating expenses, predicting revenue, and estimating the return on investment.

Expenses

A VoIP business can be started by one person, but it's more common to create a company with two or more partners who are well acquainted with the telecom market or have previous experience in similar areas. Each of the partners becomes a shareholder and takes responsibilities according to their expertise. Such a model is quite popular because the startup does not need to hire additional human resources, which take up a big part of a company's expenses. In general, the required investment and your recurring expenses depend on the business model you choose. This subsection shows the overall picture of what's needed in different business models.

Voice Traffic

The essence of VoIP business, needed for all business models. A majority of the profit is made by placing a margin on top of voice traffic and reselling it to others. The self-cost of VoIP traffic depends on the destination, quality, and other parameters which define the route. In the primary stages of a VoIP business, traffic is purchased on a prepaid basis. Once you have a solid and trusted business, you can request that your suppliers give you a credit limit. Risk and cash flow management of buying and selling VoIP traffic are especially important in a wholesale business.

Softswitch

A softswitch is the core infrastructure that is needed to run a VoIP business. It usually comes with a billing engine that allows you to add buying and selling rates and automatically calculates your profit. A softswitch can be purchased (in this case, you need to have own server or obtain a rack in the datacenter) or rented. If possible, it is always recommended to put the infrastructure in the cloud; in this way you'll have more time and energy to concentrate on those activities that help you grow your business. Price of a softswitch depends on many aspects. (You can refer back to the previous chapter, "Softswitch Selection" to learn more on this.)

VoIP Equipment

VoIP equipment—phones, ATA adaptors, routers, PBXs—is relevant for retail providers. If equipment is relatively cheap, it can be provided free for a client if they sign a long-term agreement with the service provider. VoIP equipment can also be an additional source of income for a service provider.

DIDs

DIDs are needed for a callback, calling card, call-through, and other retail services that involve using an access number or virtual number for incoming calls. DIDs can be obtained by getting a specific license that allows you to receive a number range from the regulation authority or by purchasing numbers from companies that specialize in DID aggregation and reselling services.

Softphones and Mobile Apps

Such software applications are needed for the retail service providers that deal with those who prefer to call from their smartphones or computers. Providers can choose between open source, commercial and customized applications depending on their expectations. One of the most common needs is to have a dialer integrated with the VoIP infrastructure. Investment in dialers depends on the number of users, amount of registrations, additional customization requests, and needed integration services.

Website

The website can be used not only as a virtual business card but also as a marketing and sales tool to attract clients and close deals with them. The price of building a website depends on the same factors as any other software platform or application. Usually, providers prefer to have a website that can be integrated with their VoIP infrastructure and will allow the automation of signup process. To accomplish that, it is important to have a team of developers to do needed customizations and integrations.

Integration and Customization Services

It is usually hard to find a single supplier that would be able to deliver a complete solution to run a VoIP business. Because of this many VoIP providers hire developers and integrators who can assist with the process automation and compatibility between different platforms and applications. Usually freelance developers charge based on the hourly rate, or for the total project.

Printing Services

Needed for printed calling cards and physical top-up vouchers. Such vouchers are used to increase the balance of user's SIP account. There are two ways to get vouchers or cards printed: acquire a printing machine or order printing services.

Call Shop Billing

Relevant only for a call shop business. Billing can be purchased as a standalone system (which is usually desktop-based) or as a module to the softswitch. The price of a call shop billing system is relatively small.

Gateways

Gateways are used in all cases, where there is a need to connect two different networks. Such a need usually appears in a corporate environment to connect PBXs to VoIP and in carrier-to-carrier interconnections (SS7 to VoIP, BRI/PRI to VoIP). Also, VoIP GSM gateways are a must-have product to all providers that specialize in grey termination.

SIM Bank

SIM bank is necessary only for bigger grey termination providers that have multiple VoIP gateways in different locations and prefer an easier and more convenient method of changing SIM cards.

SIM Cards and Top-ups

SIM cards and SIM card top-ups are a must-have for all providers that specialize in VoIP GSM termination. SIM cards can be blocked within a day or even within a few hours, so there should be a fast way to acquire and replace SIM cards. On the other hand, if SIM cards are not blocked often, their balance can expire. In this case, there is a need for a fast way to top-up the balance of the active SIM card.

As you see, there are no estimates of investment. However, it is important to have exact amounts in your financial calculations. To get prices, contact a few suppliers of specific services or equipment.

Income

In the simplest VoIP business model, you receive income by reselling VoIP traffic. The profit margin depends on the difference between your buying and selling rate. To be more specific, the income is received in different ways, depending on the business model.

Calling Cards

As previously mentioned, calling cards are usually divided into "white" and "black." White calling cards are a minority in the market. The biggest portion of the market consists of black cards which allow suppliers to make the highest profit. Providers of black cards put only a small margin on VoIP traffic; sometimes they even take a loss on VoIP rates. Profit in the black card business is made by applying extra fees like connection, surcharge, disconnection, daily, different increment, and others. The revenue that is received from the end-user is usually split between card distributor or reseller and calling card operator.

Retail

This is the most complex business model, and revenue streams can be different. Products and services in retail VoIP can be: voice traffic, virtual numbers, VoIP equipment (phones, ATA adaptors), and solutions (virtual PBXs, call center systems, auto-attendant solutions). Providers with a big portfolio can increase their profit margin by upselling more services to clients.

Another important aspect is the payment arrangement. Providers can choose to work on a prepaid basis (in the same way as calling cards) or postpaid (usually with a minimal one-year agreement), and charge either based on the actual consumption, or on a flat rate. In case of a flat rate, a customer gets a fixed package (e.g. 200 minutes to the USA and Canada for $10). The provider of such services benefits from unused minutes. The fewer customer calls to those destinations, the higher the profit for the provider. The normal profit margin in retail VoIP is around 20-40%.

Wholesale

Revenue is made from the price difference between selling and buying rates. Due to price wars and the low margins in wholesale, this business is cost driven. It has the leanest cost structure, low price value proposition, and maximum automation. The main criterion in wholesale is quality-price ratio. Based on this, providers can offer different rate plans (e.g. grey, standard, or premium). Revenue is dynamic and depends on:

- **Volume.** The more traffic you send, the better price you can expect.
- **Negotiation.** Bargaining is always possible, but margins are quite tight, so there is not a lot of place for a price reduction.
- **Real-time market prices.** The traffic exchange market is influenced by many factors, like regulations, wars, and politics, and is constantly changing. Because of this, wholesalers send pricing updates often, creating new opportunities to win on a specific destination daily.
- **Mixing traffic.** If a provider offers a premium plan with CLI and high quality, it is possible to get some extra profit by mixing grey traffic. Even a small portion of grey traffic can make a significant change in the revenue. However, providers should be very careful with this. If the quality drops to a certain level, the origination partner can stop sending traffic, and the revenue stream will be lost.
- **Cherry-picking.** This usually happens when a wholesaler makes a mistake or misses something in the rate sheet. Such mistakes can cause a big price gap between the real self-cost and the market price. This situation can result in large losses to the company that made the

mistake, and a high profit in a very short time for others who took a chance on the opportunity to route traffic to the low-cost destination.

Profit margins in wholesale transit are very tight, around 3-7%. Those who concentrate only on direct termination can generate bigger profits.

Grey Termination

Revenue is generated in the same way as in wholesale, except this business is usually based on one specific destination where VoIP GSM gateways are deployed. Grey traffic terminators receive revenue by selling traffic to other providers. Profit is based on the capacity capabilities (how many available ports the termination provider has) and the possibility to fill in the available ports with voice traffic. (In ideal conditions there should be a constant traffic flow to each available port.)

Let's take an example. The end-user can make an international call by using the services of the mobile operator or retail VoIP provider. If their call bypasses the mobile operator, the end-user saves 20 cents/min; the retail VoIP provider earns 20 cents/min; the VoIP GSM termination provider —3 cents/min. If the retail VoIP provider sends 20,000 minutes a day on average, the gross earnings of a grey termination provider are:

- **Daily:** $600 USD (20,000 minutes x 3 cents)
- **Monthly:** $18,000 USD ($600 USD x 30 days)
- **Yearly:** $216,000 USD ($18,000 USD x 12 months)
- **5 years:** $1,080,000 USD ($216,000 USD x 5 years)

Chapter 6.
Launching Your Service

We've covered all the important elements that you need to consider before taking a further step in the VoIP business. By now, you have:

- Understood the basic telephony concepts and the role of VoIP in the telecom market
- Chosen the service and business model that is the most suitable for you
- Selected the softswitch and additional infrastructure
- Found suppliers for voice termination, DID number, and other necessary services
- Prepared a financial plan with specific goals distributed over time

If you've completed all the above steps, the next thing is to find your first clients and test the service with them. Once you get a proof of concept, you can start to actively market and sell your services. Finally, besides running your daily operations, you'll always need to think how to adjust and improve your business to sustain and grow in the competitive VoIP market.

Sales and Marketing

In general, VoIP is a technical business, and a portion of the people who launch IP telephony services have a strong technical background. However, they often forget that to attract new clients, every business requires active sales. Customers will not be approaching your company unless you have a strong marketing and sales plan that will align to your target audience. In today's VoIP market there is a dramatic competition where many new players open and close their businesses daily; only the right strategic moves can help startups grow in such an environment.

Ideal Client Profile

Often entrepreneurs forget they can actually select their clients. This leads to the fact that some startups look for opportunities everywhere and do business with everyone who is willing to pay for their service. However, this approach may have a negative influence on the business. Note that, as a business owner, you get to choose whom you work with, and it is far better to decide this in advance.

One of the best ways to define your target audience is by creating an ideal customer profile (ICP). You may find many frameworks, methods, and templates on the Internet that will help you to complete one. When creating a profile of your ideal client, consider whether you have access to a niche market. Maybe you belong to some social group, association, closed network, or you already have clients with whom you've worked and have trusting relations. Your goal is to find a market that is not overwhelmed with VoIP offerings yet and where you can gain a competitive advantage over other providers. Your ICP will show you the picture of leads that you need to focus on.

Minimum Viable Product

Another task is to create a minimum viable product (MVP) to help you launch your services with minimal resources. Once you have developed an idea of the service that you'll be offering, you'll need to test it with your first clients. These can be your friends, acquaintances, former or current friendly customers, or prospects. A "friendly" customer does not mean someone who has only nice words for you. They should be clients who are able to be open with you and who can give both positive and negative feedback.

Once you've gathered a group of such people, encourage them to use your services on a daily basis and to share their experience. You can motivate people by providing your service at cost or even for free. Probably the best method is to select those who are passionate about the work you do. These people will be your greatest resource before your service launch.

Even if you don't make any profit from them, the experience you'll receive will be priceless. They may discover issues that you've overlooked, giving you the chance to fix them before you go to the public, resulting in a more polished final VoIP service.

Finally, if you receive positive feedback, you can post it as a testimonial on your website, giving other potential clients a chance to read reviews from other users. This helps increase trust in your company.

Sales Channels

As the beginning of this subsection, I'd like to share a story. I had a client from the USA who decided to start a retail VoIP business together with his partner. They acquired a softswitch, built a website, and spent about a year doing additional technical work. They integrated different online payment methods to a user portal, automated the signup process, customized the client interface to make it more user-friendly from their perspective, added the possibility to check calling rates online, and did a lot of additional stuff they thought was very important.

Two years passed since our first conversation, and I decided to call them and ask how things were going with their business. The client said to me, "well, things are going very well. We have finished the integration, the portal is ready to use, but we do not know what to do next."

While I was really surprised to hear this answer, it showed that there are still some entrepreneurs who are spending a lot of energy to develop the ideal product or service and who don't communicate to potential clients enough. I'm sure you won't make this mistake, but this is a story worth mentioning.

A majority of successful VoIP providers have had experience in the telecommunications or the IT sector and already had a client base before they began a telephony business. Those two factors provide a competitive advantage for companies that provide Internet access, deal with call centers, telecom operators, or offer IT services to the corporate market. The main difference between these companies and others is that they have a list of current clients who already know and trust their supplier. A database of clients is the best sales channel, but just a few VoIP startups have this resource. If you are among those who don't have this, there are some alternatives for you.

Direct Sales

All sales channels can be divided into two categories: direct and indirect. Direct is when someone from your company makes contact with a client directly. Indirect is when communication with end-users is done by your sales partners (resellers, agents, or distributors). It's better to focus on one of those channels, but you may also try a combination of both.

First, you'll need to find the way to identify and generate new prospects. You may buy a list of business contacts, use a specific resource (an online company directory, LinkedIn), or hire a person who can search for such companies. Buying a list is probably the easiest option, but it may not contain the most qualified contacts. To get a better result, search for companies in different directories and qualify them according to specific filters. If you use LinkedIn, it allows you to filter companies or people by location, industry, seniority level, interests, and other parameters. (LinkedIn is given only as an example; for your own local market you may find better alternatives.)

Once you've found a way to filter contacts, you can create a process and outsource this task for a personal assistant (also called "virtual assistant" or "VA"). There are many platforms (Amazon Mechanical Turk, Upwork, outsource.com, guru.com, peopleperhour.com, freelancer.com, Fiverr, or Zirtual) that allow you to hire virtual assistants who can do any simple job at an affordable rate.

Another way to generate prospects is by using online marketing, such as search engine optimization (SEO), AdWords, or banners. Before using any of those attraction methods, it is very important to have a clear profile of ideal clients and a good understanding of their searching behavior.

The last option that you may wish to consider is visiting trade shows and conferences to establish new contacts. (You can find a list of telecom events in the previous chapter.) However, this is the most expensive channel and is not affordable for everyone.

Once you've generated some leads, you'll need to go through the sales cycle to convert leads into paying customers. This work is usually handled by sales executives or account managers. They are responsible for closing deals with qualified leads, managing their key accounts (customers), cross-selling additional services, and seeking new opportunities. In wholesale, account managers are also named carrier relations managers. People who work in this position must have good communication, organization, and time management skills. They should be pro-active and have enough self-motivation to work efficiently under pressure with minimal supervision. If the company works in the international market, sales representatives should also have at least minimal understanding of cultural differences.

Finally, it is important to mention that if you have a good service and your target audience is big enough, you can also expect organic growth by recommendations. It is the most valuable channel that brings the best results. It might seem that recommendations do not cost anything, but actually building trust and credibility to be recommended is a hard work.

Indirect Sales

Indirect sales are achieved by establishing partnerships with complementary businesses. You may identify potential resellers in the same way as end-users; in this case, your ideal client profile should represent your sales partner. Resellers are usually companies that run a small IT or telecommunications business, or work as system integrators. VoIP is not necessarily their main source of income, but it allows expansion of their service portfolio and creates more value for their clients. In this case, it is very important to create a good channel program to motivate sales partners and allow expansion of the partner network rapidly. Dealing with resellers is great if your background is technical. You will deliver technical expertise, and your resellers will work on commercial operations and first level support for end-users. In this way you can grow voice traffic volumes faster and with fewer resources.

Here is a good example of the indirect sales channel. A company in the Netherlands began offering VoIP services as a reseller, but before long the team decided to acquire its own equipment and become a service provider because they were not happy with the quality of their supplier. Once they established their network and signed contracts with the biggest tier-1 providers in their country, they started to approach the PBX suppliers which controlled the largest portion of the telecommunications market. Their strategy was to get a certification with telephony system providers. This would allow them to be recommended as a certified service provider, guaranteed to work well with specific equipment.

Such strategy became a win-win business case. The PBX suppliers were able to introduce more alternatives to their clients. Clients were able to choose between more providers according to the quality-price ratio. Finally, this service provider got access to a big network of clients through the PBX system integrators that became their resellers. Approaching partners with an already established client base helped them to double voice traffic volume within a year.

Inbound vs. Outbound

There are two main types of marketing: inbound and outbound. Inbound is like fishing—if you want to catch specific fish, use the right kind of bait. Outbound is like hunting—choose where you pick your target and go for it. Try both options and evaluate which one works better for you.

Using inbound marketing, you need to constantly create and distribute relevant and timely content that will convert your website visitors into leads. You may use different forms of content: ebooks, educational webinars, white papers, or case studies. The most important thing is to keep track of which option works the best and expand it accordingly.

Conversion from visitor to lead is achieved when someone fills in a form on your website with their name and email address to download specific content. Once their email has been given to you, you may run a so-called "lead nurturing" process to warm up the lead. Nurturing is usually done by periodically sending interesting and relevant emails to leads. Once a lead performs a specific action, like pricing download, or requests a service trial, it's time to pass the baton to your sales team.

Outbound includes all sales and marketing methods that are meant to present products or services to potential customers without their permission. In this case, a provider actively approaches a prospective client by a different means of communication: ads, telemarketing, cold-calling, and email blasts. Most of us don't like (or hate) this option because we feel that we are disrupting a person. Moreover, it's hard to hear "no" from a prospect, and the conversion rates are relatively low.

However, outbound has some advantages over the inbound method. Outbound creates a process that can be measured and is predictable. If you have a sales funnel and know conversion rates at each stage, you can easily plan how many new leads you will need to get specific results. It gives you more control; you can hire more people or generate more leads depending in which stage of the sales funnel you have a bottleneck.

If you are interested in learning more about selling through the direct contact, read the book *Predictable Revenue* by Aaron Ross and Marylou Tyler. I had a chance to personally learn from Ms. Tyler, and that was probably the best outbound sales coaching I've ever had.

Start the Operations

Most probably at the beginning of the business you'll be handling many tasks at one time, but, in the long run, you'll learn what brings the biggest value to your company. The key is to focus on the main processes that help your business grow, and delegate or automate all the other tasks. In this section you will learn about operations that should be performed in your daily work as a VoIP service provider.

Technical Operations

A VoIP business by its nature is technological and involves the management of different types of hardware and software systems. The overall infrastructure may consist of a soft-switch, billing platform, servers, operating systems, databases, and PBX systems. On top of that, providers that have an online presence need to take care of web administration, server and domain hosting, and front-end (website, customer portal) integration with the back-end systems (softswitch, billing system, payment gateway, or shopping cards). In addition to this, there may be a need to install and configure end-user equipment or develop mobile applications.

Network and Infrastructure
If you use a cloud-based softswitch, all operations related to the server will be handled by your supplier. However, if you decide to use on-premises equipment, you'll have more responsibilities, including operations with the hardware (server), operating system, and network configurations.

The main responsibilities of managing the core infrastructure involve:

- Identifying network infrastructure problems by analyzing call logs and alerts
- Troubleshooting call quality and network issues
- Improving network design, and planning expansions if voice traffic volume is expected to grow
- Performing necessary adjustments in the softswitch: change the routing, update tariffs, and modify settings
- Interfacing with colleagues, customers, and vendors

People who work in these positions are referred to as engineers, network administrators, or NOCs. They will also be your technical experts in selecting software or hardware, and you may use their advice in negotiations with new vendors.

Maintenance of the core infrastructure may require working at irregular hours. Normally, as a service provider, you must ensure unstoppable voice service for your clients. To achieve this, you need to have 24/7 availability by creating shifts on a rotating basis or by organizing on-call support for emergencies after working hours.

Technical Support

Technical support involves providing assistance to users to fix a specific problem. The goal of technical support is to solve reported issues efficiently and in a professional manner, under conditions that were provided in the service agreement. People working in this position need a customer service focus, the ability to guide non-technical people, good communication skills, and high quality work ethics.

If your company is small, support can be provided by the same person who manages the system administration. But in larger organizations there are usually a few levels of technical support. The first level involves answering phones, helping with basic questions, and forwarding more serious tasks to another level. The second level is responsible for the actual solving of technical problems, and, if a solution requires changes in global infrastructure, passing the problem to the third level. The third level itself takes care of the global system administration.

Technical support may be provided by phone, email, online chat, or a support system where users can create so-called "tickets" or "incidents." Such systems allow a review of the problem and the tracking of its progress and are also easier for managing support efficiency and evaluating the most frequent technical problems.

Web-related Operations

Most retail VoIP providers have an online presence, and their ultimate goal is to completely automate the service-buying process and improve the end-user's experience. To achieve this, they need to take care of web administration and ensure interoperability between the website and other systems.

Whoever manages web-related operations must ensure that the website operates correctly. This may involve work with the web server hosting system, content management system (CMS) (such as WordPress, Drupal or Joomla), and general revision of web pages. Most of those tasks do not require any programming knowledge and can be completed through the intuitive interface. They usually do not take a lot of time and can be done by the same person who manages technical operations in your company. However, if your website requires integration with third party systems, or if you need to create your own customized user portal, you'll need to consider hiring a freelance developer.

End-user Equipment and Applications

If you plan to provide end-user equipment, such as ATAs, IP phones, or PBXs for your clients, you'll need to take care of installation and configuration (unless you have technology partners for that). If possible, it is recommended that you deliver pre-configured equipment with auto-provisioning enabled. (Please consult with your softswitch or PBX supplier to see if this feature is supported.) This ensures seamless connectivity between the device and the service provider's infrastructure. The auto-provisioning function allows end-users to simply plug the phone or ATA in to the Internet and start calling straight away. Moreover, this function makes it much easier to run software updates and troubleshoot problems.

However, not all the work can be done remotely. You should expect that provisioning of end-user services may also involve visiting your clients to resolve various incidents. Your engineers should be able to work on client premises to run diagnostics and apply needed adjustments in configuration settings.

Billing and Invoicing

Billing is either done by a standalone piece of billing software or a softswitch (if it incorporates billing functions). The overall process covers many areas, such as rate management (applying tariffs, discounts, and taxes), generating and sending invoices, credit and balance control, CDR reconciliation, disputes and adjustments, report generation, and processing payments. Such operations can be done automatically or by a responsible person.

Rate management is a more relevant task for carrier relations managers who work in the wholesale transit business. They need to deal with many rate sheets in different formats provided by their current and potential suppliers. Because of this, they need to have an efficient and easy-to-use tool to handle this process fast, in order to focus on their main activity—sales.

Another task, usually handled by accountants, is invoice generation. The process of generating invoices is usually automated and needs to be done each month or week, depending on the billing period. After invoices are ready, the accountant may review them manually and, if necessary, apply adjustments before sending them to clients. Once invoices are delivered, it is important to make sure that payments are made on time, according to agreed conditions. Usually each provider has its own follow-up and reminder strategy to handle this. Clients may pay using an online payment processor or a wired bank transfer. In the first case, their balances are automatically updated, and invoices are marked as paid. In the second case, payments should be added manually, unless there is an integration between the bank account and billing system.

The last billing-related task is CDR reconciliation. This is done in case there is a dispute between client and supplier. CDRs should be recalculated and compared with different billing systems. After discrepancies are found, necessary adjustments should be reflected on the invoice.

Grow Your Business

To be a competitive provider, you'll need to have realistic goals and a team of professionals who will help you perform all necessary tasks. As a startup, be always ready to change the business direction completely and adjust your primary business plan. Here are a few tips for growing a successful VoIP business.

Organize, Prioritize, and Focus

Organization helps you to be aware of the tasks that must be done under different levels of priority. Focus only on the important activities that move your business forward. All remaining tasks that are not important should be automated, delegated, or declined. There are many methods to organize and prioritize. The one personally used and liked the most is called the "Eisenhower matrix" or "Eisenhower box." It helps to evaluate urgency and importance. Items may be placed at more precise points within each quadrant (see the matrix below).

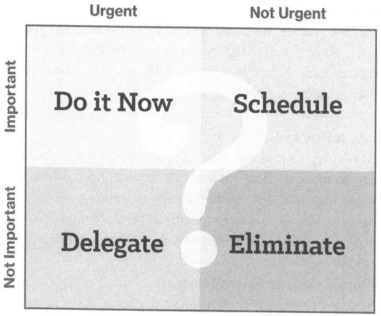

Figure 6.01. Eisenhower matrix.

Know Your Competitors

You need to know your top competitors and to keep track of their activities. This does not necessarily mean you must respond to each of their actions. But you can learn new things from your competitors and consider implementing something similar, just more improved, in your business. It is also important to note that many VoIP startups assume that Skype and Viber are their top competitors, but in most cases that is not true. Try to compare your services with other providers that work in the same market on a similar scale as you.

Understand the Risks

According to statistics, only 18% of first-time startups become successful. However, a majority of entrepreneurs are too ambitious at the beginning of their business and think their idea is different and will be a success. Calculate the risks and ask yourself about the downsides of the business you are going into. Another suggestion is to think about the worst-case scenario before starting the business. This helps to prepare morally before taking the risk.

Be Creative and Innovative

Always try to look for new ways to improve your business and to make it stand out from the competition. Admit mistakes, recognize that you don't know everything and try to be open to new ideas, suggestions from your team and requests from your clients. Creativity moves the business forward.

Provide Good Service and Listen

Business revenue comes from your clients. This is quite a strong argument for constantly improving your services and customer support. Listen and be aware of their needs. In the long run, you may have many different requests that you won't be able to handle. Always remember to focus on the issues that are a must-have for your customers or that give you a competitive advantage.

Keep the Pareto Principle in Mind

It is also known as the 80–20 Rule: 80% of the effects come

from 20% of the causes (20% of the clients bring 80% of your income). Again, this allows you to focus on the top clients, the key activities, and to concentrate on business growth.

Hire Smart People

Once your business grows and you see that you cannot handle everything on your own, start hiring people. If you want the business to be scalable, hire creative, accountable and smart people who will carry out some activities better than you. Delegate work to your colleagues and trust them; don't keep control of all operations. For the simple tasks, you can use personal assistants.

Final Word

Starting and running a successful VoIP business is challenging, but in the end you will become your own boss and have the freedom to make your own decisions. Moreover, you have a greater potential to earn more. Success requires discipline, long-term focus, consistency, and flexibility due to changing environments. It does not happen overnight. For all these reasons, I wish you great patience in your VoIP business, and I hope it will be a successful journey for you!

I've tried to add the information most relevant in this guide, but there is always room for improvement. I'd really like to hear your opinion of this book, so if I've missed something, let me know! You can share your feedback (or criticism) and ask any questions by contacting me at http://www.runvoipbusiness.com.

About the Author

Vilius Stanislovaitis has been working in the telecommunications industry for nine years. In his most recent position, he helps startups and service providers manage a VoIP business more efficiently. During his career, he has worked with thousands of startups and small VoIP service providers from 90 countries.

Frequent questions from entrepreneurs interested in starting a VoIP business encouraged the author to begin research on this topic. There were no books available, which would explain how to start different business models based on VoIP technology. This inspired the author to create a guide that would give a basic understanding about starting a VoIP business, for everyone curious about this evolving and expanding industry.

Acknowledgements

Many thanks for Mindaugas Kezys, founder of Kolmisoft. I've learned many things about VoIP from him, and without Mindaugas, this book might not have been started at all.

Prior to its release, the manuscript of this book was reviewed by 185 people; thanks to all of them. I especially appreciate the extraordinary input to the content of the book by Leonid Karlinsky. All of his suggestions were added to this book and thanks to him, you can now read a book which was carefully reviewed by a successful VoIP business owner. Also thanks to Alessandro Valenti, Boris Vutov, and Giedre Rubyte. You helped to improve small things that matter.

Part of the book's financing was done through Kickstarter. Thanks to all 200 backers from 36 countries who supported the release of this book! You've done a great job and helped my long-term goal come true! Without you, the book wouldn't have been professionally edited, designed and printed. Special thanks for Alessandro Marzini, Marylou Tyler, Bas Kooijman, Mindaugas Kezys, Dhesigan Naidoo, Rohin Ravindran, 4AllBusiness team, Kalyan Kumar Pasumarthy, James Murdock, Peter Imeokparia, Simon Kusugh, Marcus Joyce, Ron Byer, Hubert Mickael, Sam Shukr, Bruno Fernandes, and Omid Omidvari.

Apart from the people who provided direct assistance, I'd particularly like to mention Guy Kawasaki, author of the book *APE: Author, Publisher, Entrepreneur*. His book was an irreplaceable guide for me during this journey of self-publishing.

Glossary

[A]

Access number
A regular number that is used to access services of a VoIP provider to make international calls. Access numbers are widely used in calling cards, callback, and call-through services. Access numbers are also known as DIDs or virtual numbers.

Account manager
A position in wholesale VoIP that is similar to a sales manager in any other company. Account managers are responsible for finding clients and suppliers to sell and buy "routes" and to keep up relations with them (follow up regarding payments or announcements of new offers).

Add-on (module)
A piece of software that enhances the main software application and cannot run independently.

Analog telephony adapter (ATA)
A device for connecting traditional analog telephones, fax machines, or other devices to a VoIP network.

Answer-seizure ratio (ASR)
A quality measurement in telecommunications which shows the percentage of answered telephone calls with respect to the total call volume.

Application programming interface (API)
An interface by which an application program accesses operating systems and other services. It is used for building software applications and integrating them to each other.

Authentication
The process of verifying that a user is who they claim to be. The user supplies a username and password; then the system accepts these as inputs and verifies that the user is valid and has access to the system.

Authentication, authorization, and accounting (AAA)
A framework for intelligently controlling access to system resources, enforcing policies, auditing usage, and providing the information necessary to bill for services.

Authorization
The process of verifying what an authenticated user is allowed to do.

Auto dialer
A piece of software that automatically dials telephone numbers and plays a recorded message.

Automatic number identification (ANI)
A feature of a telecommunications network for automatically determining the origination of a telephone number. It saves time for a client who does not need to enter a PIN or other identification code to call. The customer is identified automatically by Caller ID.

Auto-provisioning
A process of applying or running a service (registering a device, updating a device, or changing global settings) automatically without requiring human intervention.

Average call duration (ACD)
A quality measurement in telecommunications that reflects an average length of calls.

A-Z
A term referring to all the global destinations from letters A to Z. It is used to describe routes (A-Z routes), rate sheets (A-Z rates), and other services, offered by a wholesale VoIP provider.

[B]

Bandwidth optimization
A process of reducing Internet bandwidth consumption.

Bandwidth optimizer
A solution that allows reducing Internet bandwidth consumption.

Basic Rate Interface (BRI)
A telecommunications interface standard used on an ISDN that provides two bearer channels (B channels) and one data channel (D channel). The B channels are used for voice or user data, and the D channel is used for any combination of data, control, and signaling.

Bilateral agreement
A contract involving mutual promises for each party/both parties. This term in VoIP transit business means that each provider can be both—client (originator) and supplier (terminator).

Billing system
The system that performs the billing processes of telephony business (rate management, invoicing, tariff plans, and payment processing).

Bring your own device (BYOD)
The policy of permitting customers to use personally-owned devices (IP phones, PBXs, and mobile phones) for a service supplied by the provider.

[C]

Call capacity
A measure determining the maximum traffic volume and the highest load that can be handled by a system or sent through a channel at specific conditions.

Call detail record (CDR)
A record of a telephony event that documents the details of a call that passes through the system or the device.

Call forwarding
A telephony feature which redirects a call to another destination (for example, a mobile telephone, voicemail, or an external number).

Call legs (Leg A and Leg B)

A term, used to describe active pieces of a call (Leg A and Leg B). In the strictest technical sense, the Leg A represents ingress to the switch while the Leg B represents egress from the switch. In most cases this means that the originator of the call is the Leg A and the recipient of the call is the Leg B.

Call origination

See: Origination.

Call shop

A business providing on-site access to telephones for making long-distance calls.

Call termination

See: Termination.

Callback

A service where the originator of a call is immediately called back as a response to the call trigger.

Caller ID (CID)

A telephone service that transmits a caller's number to the called party's telephone equipment during the ringing signal or when the call is being set up, but before the call is answered.

Caller ID spoofing

A practice of causing the telephone network to indicate to the call receiver a different calling number from the original one.

Calling card
A card used to pay for telephone services (usually long-distance calls).

Calling line identification (CLI)
A telephone line from which a telephone call originates. This term is also used to describe telecom routes (see: CLI route).

Call-through
A service (usually for long-distance calls) that involves dialing the access number followed by the destination number.

Carrier relations manager
See: Account manager.

CDR reconciliation
The process of matching CDRs from two or more different sources to find discrepancies among them.

Channel
A transmission medium in telecommunications that is used to convey an information signal.

Circuit switching
A methodology of a telecommunications network in which two network devices establish a dedicated communications channel (circuit) through the network before they communicate.

Class 4
A term used to describe the class of a switch (softswitch).

Class 4 switches are used in wholesale for carrier-to-carrier interconnections.

Class 5
A term used to describe the class of a switch (softswitch). Class 5 switches are used in retail to deliver services for the end-users and subscribers.

CLI route
A type of telecommunications route, where Caller ID is saved (opposite to NON-CLI or NCLI).

Closed-source software
A proprietary piece of software with source code that is not available in public.

Codec
Used to convert an analog voice signal to a digitally encoded version.

Competitive local exchange carrier (CLEC)
A telecommunications provider competing with other already established carriers, generally the incumbent local exchange carrier (ILEC). The term is mostly used in the USA.

Concurrent calls (CC)
Calls that are taking place at the same time.

[D]

Database (DB)
An organized collection of data.

Dial plan
Specifies how to interpret digit sequences dialed by the VoIP user, and what telephony features should be activated in this case.

Dial telephone
A telephone that allows dialing (earlier you could not dial; only pick up the phone).

Digital
Data that is represented using discrete (discontinuous) values.

Digitization
Converting analog signals into binary bits for storage and transmission.

Direct inward dial (DID)
A standard outside number that allows reaching a customer's PBX system, individual extension, application, or external number.

Direct termination
A telecommunications service, provided through the direct network connection with the final user or the PSTN.

Dispute
A record of a disagreement raised by a customer or partner, typically in reference to a charge.

Distributed environment
An implementation where network elements are deployed separately from each other.

Dual-tone multi-frequency signaling (DTMF)
The signal to the phone company that is generated when user presses an ordinary telephone's touch keys.

[E]

E1
The European format for digital transmission. It contains 32 time slots: one is used for signaling, another for synchronizing, and the remaining thirty for traffic (voice/data).

Encryption
The process of encoding digital data in such a way that only authorized parties can read it.

Extension
An additional telephone set connected to the same telephone line as another set or sets.

[F]

File transfer protocol (FTP)
A standard network protocol used to transfer computer files from one host to another host.

Flat rate (flat-fee)
A pricing structure that charges a single fixed fee for a service, regardless of usage.

[G]

General Public License (GPL)
The free software license that guarantees end-users (individuals, organizations, companies) the freedom to use, study, share (copy), and modify software.

Global System for Mobile Communications (GSM)
A standard to describe protocols for second-generation (2G) digital cellular networks used by mobile phones.

Graphical user interface (GUI)
A program interface that takes advantage of the computer's graphics capabilities to make the program easier to use.

Grey route
A term which defines a route that is legal for one country or the party on one end, but illegal on the other end.

[H]

H323
The protocol used for audio and video communication sessions on any packet network.

[I]

Inbound marketing
Promoting a company through blogs, podcasts, video, ebooks,

newsletters, white papers, social media marketing, and other forms of content marketing which serve to attract customers.

Incumbent local exchange carrier (ILEC)
A local telephone company that held the regional monopoly on landline service before the market was opened to competitive local exchange carriers.

Integrated Services Digital Network (ISDN)
A set of telecommunication standards for simultaneous digital transmission of voice, video, data, and other network services over the traditional circuits of the PSTN.

Interactive voice response (IVR)
A technology that allows a computer to interact with humans through the use of voice and DTMF tones input via keypad.

Inter-Asterisk eXchange (IAX)
A communications protocol native to the Asterisk PBX software, and supported by a few other softswitches, PBX systems, and softphones.

Interconnection
A virtual or physical linking between networks of two or more carriers.

Interconnection rate
A rate at which a provider agrees to terminate calls inside its network.

Internet protocol (IP)
The principal communications protocol in the Internet protocol suite for relaying datagrams across network boundaries.

Internet service provider (ISP)
An organization that provides services for accessing, using, or participating on the Internet.

Internet telephony
See: Voice over Internet protocol (VoIP).

Internet telephony service provider (ITSP)
A company that offers telecommunications services based on VoIP that are provisioned via the Internet.

Invoice
A legal document demanding payment for one or more product instances, and their associated events, for a single account.

IP address
A numerical label assigned to each device (e.g., computer, IP phone) participating in a computer network that uses the Internet protocol for communication.

IP PBX
A PBX used in packet networks.

IP phone
A phone used in packet networks.

IP telephony
See: Voice over Internet protocol (VoIP).

[J]

Jitter
The deviation from true periodicity of a signal in telecommunications.

Jitter buffer
A shared data area where voice packets can be collected, stored, and sent to the voice processor in evenly spaced intervals.

[L]

Last mile
The final leg of the telecommunications networks delivery to end-users.

Least cost routing (LCR)
The process of selecting the path of outbound communications traffic based on cost.

Local number portability (LNP)
Portability of local numbers (see: Number portability).

[M]

Mean opinion score (MOS)
A quality measure in telephony networks used to obtain the human user's view of the quality of the network.

Media server

A server which processes and generates media streams.

Mobile dialer (mobile app)

An application for smartphones, used for incoming and outgoing calls.

Mobile number portability (MNP)

Portability of mobile numbers (see: Number portability).

Multi-tenancy

A software architecture in which a single instance of a software runs on a server and serves multiple tenants (e.g. multitenant PBX).

[N]

NAT traversal (NAT-T)

A general term for techniques that establish and maintain Internet protocol connections traversing network address translation (NAT) gateways that break end-to-end connectivity.

Network address translation (NAT)

A methodology of remapping one IP address space into another by modifying network address information in Internet protocol (IP) datagram packet headers while they are in transit across a traffic routing device.

Network latency (delay or latency)

A quality measure in telecommunications that shows the delay from input into a system to desired outcome.

Network operations center (NOC)
The technical personnel responsible for monitoring telecommunications networks, analyzing problems, troubleshooting, and communicating with site technicians.

Network topology
The arrangement of the various elements of a telecommunications network.

Number portability
The process that allows a customer to keep their telephone number when they change their operator.

[O]

Off-net
Outside the provider's network.

On-net
Inside the provider's network.

On-premises
Software or hardware that is installed and run on computers on the premises (in the building) of the person or organization, rather than at a remote facility (virtual environment, cloud, hosted).

Open source
A development model which promotes universal access via a free license to a product's source code, and universal redistribution of it (including subsequent improvements to it by anyone).

Origination
The collecting of calls initiated by a calling party on the PSTN, and handing off the calls to a VoIP endpoint or to another exchange or telephone company for completion to a called party.

Originator
A telephone company that provides call origination services.

[P]

Packet loss
A quality measure in telecommunications that occurs when one or more packets of data traveling across a computer network fail to reach their destination.

Packet-switched
A digital networking communications method that groups all transmitted data into suitably sized blocks called "packets" that are transmitted via a medium that may be shared by multiple simultaneous communication sessions.

Payment gateway
An e-commerce application that authorizes credit card payments for e-businesses.

Payphone
A coin-operated public telephone, often located in a telephone booth or a privacy hood, requiring prepayment by inserting money (usually coins) or by billing a credit or debit card.

Personal identification number (PIN)
A numeric password shared between a user and a system, that can be used to authenticate the user to the system.

Pinless
The VoIP service that allows convenience making calls without entering a PIN number.

Plain old telephone service (POTS)
A telephone service employing analog signal transmission over copper loops.

Portal (client/user/reseller portal)
See: Graphical user interface (GUI).

Post dial delay (PDD)
The time between the start of the call and the moment that the phone of the called party starts ringing.

Predictive dialer
A piece of software which dials a list of telephone numbers and connects answered dials to people making calls, often referred to as agents.

Presence
A feature of a phone or PBX that makes it possible to locate and identify a device wherever it might be, as soon as the user connects to the network.

Primary Rate Interface (PRI)
A telecommunications interface standard used on an ISDN

that provides 23 bearer channels (B channels) and 1 data channel (D channel). The B channels are used for voice or user data, and the D channel is used for any combination of data, control, and signaling.

Private branch exchange (PBX)

A telephone exchange or switching system that serves a private organization and performs concentration of central office lines or trunks and provides intercommunication between a large number of telephone stations in the organization.

Programming error (software bug)

An error resulting from bad code in some program involved in producing the erroneous result.

Proprietary software (commercial software)

A piece of software licensed under exclusive legal right of the copyright holder with the intent that the licensee is given the right to use the software only under certain conditions and restricted from other uses, such as modification, sharing, redistribution, or reverse engineering.

Protocol conversion

The process of converting the protocol of one device into a protocol suitable for another device to achieve interoperability.

Proxy

See: SIP proxy.

Public switched telephone network (PSTN)
The aggregate of the world's circuit-switched telephone networks that are operated by national, regional, or local telephone operators, providing infrastructure and services for public telecommunications.

[Q]

Quality of service (QoS)
The overall performance of a telephony network, particularly the performance seen by the users of the network.

[R]

Real-time Transport Protocol (RTP)
A network protocol for delivering media stream (audio and video) over IP networks.

Redundancy
The duplication of critical components or functions of a system with the intention of increasing reliability.

Remote Authentication Dial-In User Service (RADIUS)
A networking protocol that provides centralized authentication, authorization, and accounting (AAA) management for users who connect and use a network service.

Reseller
A person or organization that resells the products or services of their provider. Resellers usually do not have their own infrastructure and work under the provider's brand or as a white-label (see: White-label service).

Route

A term used to describe the product of voice traffic (usually, international). There are different types of routes: grey, white, direct, CLI, NON-CLI, and more.

RTP server

See: Media server.

R-value

A quality metrics in telecommunications that uses mathematical formula incorporating network latency, jitter and packet loss, and grades on the scale of 1 (unintelligible) to 100 (very clear).

[S]

Search engine optimization (SEO)

The process of affecting the visibility of a website or a web page in a search engine's unpaid results, often referred to as "natural," "organic," or "earned" results.

Secure Real-time Transport Protocol (SRTP)

A profile of RTP (Real-time Transport Protocol), intended to provide encryption, message authentication and integrity, and replay protection to the RTP data in both unicast and multicast applications.

Session border controller (SBC)

A device regularly deployed in VoIP networks to exert control over the signaling and the media streams involved in setting up, conducting, and tearing down telephone calls or other interactive media communications.

Session initiation protocol (SIP)

A communications protocol for signaling and controlling multimedia communication sessions.

Session Traversal Utilities for NAT (STUN)

A standardized set of methods and a network protocol to allow an end host to discover its public IP address if it is located behind a NAT.

Signaling System No. 7 (SS7)

A set of telephony signaling protocols, used to set up and tear down most of the world's public switched telephone network (PSTN) telephone calls.

SIM dialer

An attachment to a SIM card that can store a few access numbers.

SIP client (SIP user agent)

A hardware device (IP phone, ATA) or software (mobile dialer, softphone) that allows peer-to-peer calls to be made using SIP protocol.

SIP proxy (SIP proxy server or SIP server)

One of the main components of a telephony system (softswitch, IP PBX) that is used to perform many of the call set-up functions, such as user registration, authentication and authorization, or implementation of call routing policies.

Softphone

A software program for making telephone calls over the Internet using a general purpose computer, rather than using dedicated hardware.

Softswitch

Short for software switch (see: Switch).

Standalone

Hardware or software that works on its own and is not a part of some bundled solution.

Subscriber identity module (SIM)

A removable module that is inserted into a mobile terminal (mobile phone, VoIP GSM gateway). The card contains all subscriber-related data, such as access numbers, service details, and memory for storing messages.

Switch (switching system)

A central device in a telecommunications network that connects telephone calls from one phone line to another, across a telecommunications network or the public Internet.

Switch partitioning

A feature that allows dividing a switch into separate, autonomous accounts.

Switchboard

A telecommunications system used in the public switched telephone network or in enterprises to interconnect circuits of telephones to establish telephone calls between the subscribers or users, or between other exchanges.

Switchboard operator

People responsible to connect calls by inserting a pair of phone plugs into the appropriate jacks of a switchboard.

Switchless resellers

Resellers of telecommunications services that do not have their own switch (equipment).

[T]

T1

The most commonly used digital transmission service in the United States, Canada, and Japan. It consists of 24 separate channels.

Telephone exchange

A telecommunications system used in the PSTN or in enterprises. For corporate use, a private telephone exchange is often referred to as a "private branch exchange" (PBX).

Telephony

The field of technology involving the development, application, and deployment of telecommunications services for the purpose of electronic transmission of voice, fax, or data, between distant parties.

Telephony interface card (telephony card)

A card that converts legacy signaling and media into another telephony format.

Termination
Routing of calls from one provider to another or connecting calls to the end point (end-user).

Terminator (termination operator)
Telephone company that provides call termination services.

Tier
A term used in telecommunications which describes levels of carriers.

Time-division multiplexing (TDM)
A method of transmitting and receiving independent signals over a common signal path by means of synchronized switches at each end of the transmission line so that each signal appears on the line only a fraction of time in an alternating pattern.

Toll-free number
A telephone number that is billed for all arriving calls instead of incurring charges to the originating telephone subscriber.

Traditional telephony (conventional/legacy telephony)
A term used to describe circuit-switched telephony (PSTN, POTS).

Transcoding
The process of converting codecs from one format to another.

Transcoding cards
The card that performs the transcoding process.

Transfer
A feature in a telephony device (phone or PBX) that allows the forwarding of a call to another internal or external number by pressing a "transfer" button.

Transit operator
A telephone company that provides VoIP traffic transit services.

Triple-play
A term for the provisioning over a single broadband connection of broadband Internet access, television, and telephony services.

Trunk
A line or link that can carry many signals at once, connecting two or more exchanges, switching centers or other devices or communications systems.

[U]

Unified communications (UC)
The integration of real-time, enterprise, communication services such as instant messaging, presence, voice (including IP telephony), mobility features, audio, web and video conferencing, fixed-mobile convergence, desktop sharing, data sharing, and more.

[V]

Virtual number
See: Direct inward dial (DID).

Virtual PBX (cloud PBX)

A service where the call platform and PBX features are hosted at the service provider location.

Virtual private network (VPN)

A network that uses a public telecommunications infrastructure, such as the Internet, to provide remote offices or individual users with secure access to their organization's network.

Voice broadcasting

A process of sending pre-recorded voice messages to a list of call recipients.

Voice over Internet protocol (VoIP)

The technology that describes routing voice conversations over the Internet or any other packet-switched network.

Voice traffic

A commodity in the telecom business which represents voice transmission services to a specific destination.

Voicemail

A method of storing voice messages electronically for later retrieval by intended recipients.

VoIP GSM termination

A service where calls are originated in VoIP and terminated in the GSM network (see: Grey route).

VoIP GSM/PSTN gateway
The gateway that converts VoIP to GSM/PSTN and vice versa.

VoIP provider
See: Internet telephony service provider (ITSP).

VoIP traffic
See: Voice traffic.

[W]

Web phone
A web-based softphone that allows making calls directly from a website or other web solution (like user portal).

Web Real-Time Communication (WebRTC)
A standard that supports browser-to-browser applications for voice calling, video chat, and other functions without the need of either internal or external plugins.

White route
A route that is legal for both origination and termination parties.

White-label service
A service produced by one company (provider) that other companies (resellers) rebrand to make it appear as if they had made it.

Wireless

A telephone service that uses electromagnetic waves to carry a signal, rather than sending it via cable or wires.

Wireless Internet service provider (WISP)

An Internet service provider with a network based on wireless networking.

9 786094 088308